BASIC HOME REPAIR & MAINTENANCE

BASIC HOME REPAIR & MAINTENANCE

An Illustrated Problem Solver

TERRY MEANY

Guilford, Connecticut

An imprint of The Rowman & Littlefield Publishing Group, Inc.
4501 Forbes Blvd., Ste. 200
Lanham, MD 20706
www.rowman.com

Distributed by NATIONAL BOOK NETWORK

British Library Cataloguing in Publication Information available

Library of Congress Cataloging-in-Publication Data available

ISBN 978-1-4930-5927-0 (paper : alk. paper)
ISBN 978-1-4930-5928-7 (e-book)

∞™ The paper used in this publication meets the minimum requirements of American National Standard for Information Sciences—Permanence of Paper for Printed Library Materials, ANSI/NISO Z39.48-1992.

Acknowledgments

It's the author's name that shows up on the book's cover, but there are editors, proofreaders, and more unseen professionals in the background who bring the writer's musings to life. I want to thank Maureen Graney at Globe Pequot Press for offering me the opportunity to write this book, knowing she had any number of other authors from which to choose. I also want to thank Katie Benoit for her editorial prodding towards producing a finished manuscript. Finally, thank you to whoever created the "delete," "cut," and "paste" computer features without which I never, ever would have finished this book.

CONTENTS

INTRODUCTION

New is very appealing: new clothes, new electronics, or a new car. New usually means shiny clean, everything in working order, no worries or concerns; whatever this new thing is, you can simply use it and enjoy it. Eventually, the new clothes need cleaning or mending, the computer balks at a strange software program, or that imported sedan's "check oil light" goes on. The honeymoon isn't completely over, but the relationship is going to require a bit more attention now. With a home, especially an older one, the relationship might require some extensive counseling, as your furnace, roof, plumbing, and paint, just to name a few, all demand your attention.

Let's establish one truism up front: No one likes to do home maintenance. We would all like our appliances, floor finishes, garage doors to work perfectly and look great forever despite using (or abusing) them day after day. Without a crew of handy guys and girls discreetly hidden away in the servants' quarters, the maintenance and repair tasks fall on you. And you will not always feel comfortable with the jobs you'll face.

In *The Far Side* cartoon entitled "An Elephant's Nightmare," an elephant sitting in front of a grand piano at a fully packed concert house thinks, "What am I doing here? I can't play this thing! I'm a flutist, for crying-out loud!" If you're unfamiliar with sink drains, electrical circuits, entry door locks, and plaster repair, you might feel like this out-of-place pachyderm flutist. Your main tool might be a phone book to call unknown contractors and businesses specializing in repairs to fix your problems.

But this approach prompts other questions: Whom do you call? Do you even need a professional? Do you have the time and money to hire and wait for a plumber? What happens if a sudden freeze bursts a water pipe in the middle of the night? Knowing enough to handle most problems with some skill and success can turn a major predicament into a minor inconvenience.

You don't have to be an expert in plumbing, carpentry, tile setting, or any other construction trade to work on your house. Your results don't have to be perfect, but knowing enough to quickly solve or temporarily fix a problem is a major benefit. With nationwide

home center chains, neighborhood and regional hardware stores, books, home repair TV shows, and online help, anyone can tackle painting a bedroom or stopping a dripping faucet. All you need to get started is a question-and-answer session with yourself.

Is it important to do this repair, or can I put it off?

Chipped paint or scratched wood floors look unsightly, but they can be left alone as long as you're willing to live with their appearances. A loose and overflowing gutter should get attended to before it causes more damage. Ignore regular furnace maintenance, and you'll regret it when the furnace stops running on a cold winter day. Some jobs, such as touching up the exterior paint, can be let go during a warm, dry summer but should be taken care of before the fall weather sets in.

Can I do this job myself?

We might like new things, but we don't always like new experiences that take us outside of our comfort zone of skills. Maintaining a home exposes us to a lot of new experiences we might prefer to ignore or even flee from. This book will help you face them with the confidence to figure out how to fix the problems. Practice leads to improvement and more comfort with tools and techniques. It doesn't have to lead to perfection, unless you find you really enjoy building fences or hanging drywall. Often enough, you'll be surprised at what you can accomplish.

Should I do this job myself?

If you have one bathroom, it's a Friday afternoon, and you're expecting weekend company, it probably isn't the time to replace the inner workings of your only toilet. Call a plumber. Smell gas around your furnace or water heater? Call the gas company. Afraid of heights, and

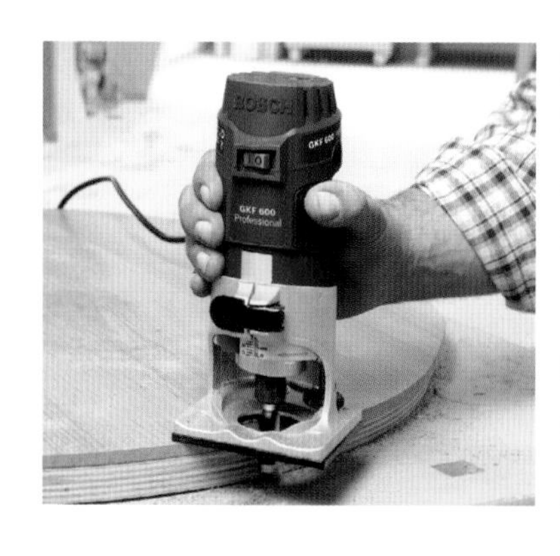

your roof is missing a few shingles after a recent wind storm? Contact a roofer.

Your kitchen has seen better days and needs a new paint job? It might be tedious working around all the cabinets and fixtures, but have a go at it. If nothing else, you'll gain a new appreciation for the painting trade and a different perspective on the notion that "anyone can paint." Your dryer suddenly stops drying, but the drum keeps turning? It's probably a heating element and something you can replace and save on an appliance repair technician's service call.

In other words, know your ***limitations,*** but don't underestimate what you ***can*** do.

Do I have the time to do this?

Even if it's a job you can do, if it doesn't fit your schedule, you might have to hire it out. Some of us believe we should fix everything ourselves, but if it isn't practical, the repair will either go undone or will be done in a hurry, either of which can lead to some regret. Home maintenance is about time and money management as well as the repair work itself. Even people who do repairs hire some of their own house work out to other contractors.

Is it worth paying someone to do this work?

Contractors are not inexpensive. Each has to account for overhead, materials, taxes, transportation, losses on some jobs, and profit the same as you or your employer does in your line of work. Repairs are normally paid for with after-tax dollars, meaning if a contractor is charging you \$50 an hour, you would have to earn over \$60 an hour before taxes to pay the bill. Doing the work yourself keeps the money in your wallet.

However, if it takes you five times as long to do the work and three car trips to a home center for parts and the results are only so-so, then, aside from the learning experience, this job might not be much of a cost saver. In some cases, you won't find this out until you've completed the work. Carefully consider the cost and time advantages and disadvantages before determining to take on a repair.

Maintenance can be divided into two categories: preventative and immediate

Preventative maintenance is the unexciting, ongoing routine tasks that keep your home running smoothly. This includes regular gutter cleanings, paint touch-ups, and furnace inspections. What else falls under preventative maintenance? Cleaning moss off the roof, checking the chimney for creosote, replacing worn weather stripping, and sealing off any outside holes that could be used by various critters to crawl inside your home are all part of a regular to-do list. You don't have to do anything on this list, but ignoring these tasks can lead to bigger tasks later.

Preventative maintenance helps put off most *immediate maintenance.* A broken washing machine hose, a chimney fire, and squirrels nesting in your attic are examples of immediate or even emergency maintenance. You can't foresee every possible repair, but many can be headed off. An ounce of prevention really is worth a pound of cure when it comes to your home.

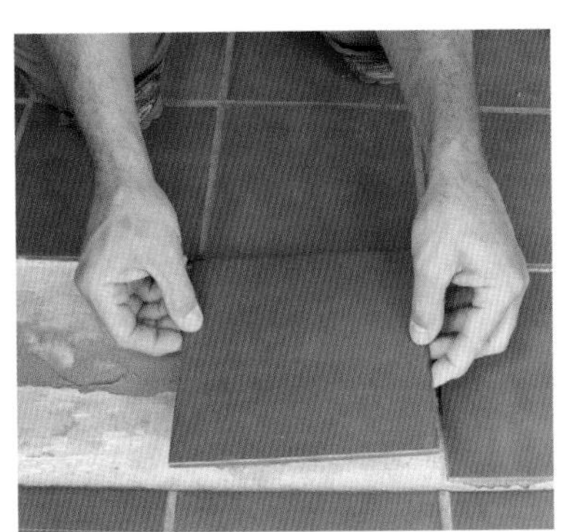

How much time do you need to devote to home maintenance? It depends on how many people live in your home and how they act in it. A single person who is away on business half the year will probably spend more time dusting than cleaning shower drains. A family of six amateur acrobats who practice in the living room and share their home with a small menagerie of animals allowed to run loose inside is a different story! Staying ahead of your maintenance is easier in the long run than falling behind.

This book will give you the tools and guidance to take a crack at many of your home repair maintenance and problems without feeling overwhelmed. And when you're not overwhelmed, you'll enjoy your home even more.

HOW TO AVOID PROBLEMS

Schedule regular maintenance check-ups to avoid any problems in the first place

"A stitch in time saves nine" is an American proverb that harks back to the times when we still repaired torn clothes instead of tossing them, but the sentiment remains true. Catch a problem early, and it usually stays manageable. Ignore it, and it becomes a bigger problem at a bigger cost. Most of the systems and parts that make up your home are reliable, at least when they're new. Time passes, and malfunctions show up. They're to be expected. You can run a dryer or turn a key in a lock only so many times before something gives way. You can't prevent every problem, but you can slow some up, head others off, and be prepared when they happen.

Open Toolbox

- Tools are one of the best investments you can make.
- For the most part, good hand tools will last a lifetime or certainly many years, far more than just about anything else we buy.
- You don't need one of every tool made, just enough of the basics to do most jobs—the rest you can get as needed.
- You also don't need the very best tools—decent quality for normal maintenance is more than adequate.

Measuring the Job

- Accurately measuring can simplify a job, and not measuring can complicate a job—simplify is much better.
- Estimating the size needed for a repair task is fine as a rough guideline, but measuring means fewer trips to the store, a closer cost estimate, and less wasted material.
- If you measure twice and still aren't comfortable cutting or drilling, then measure again until you are comfortable.
- Don't depend on eyeballing anything—use a measuring tape.

Preventative maintenance isn't exactly fun, but it beats the alternatives, such as a leaking roof during a rain storm or non-functioning furnace in the dead of winter. Scheduling regular furnace inspections, replacing missing roof shingles, touching up peeling paint, and knowing when to replace a dying dishwasher helps you to control your home environment and keep it safe and inviting. It's best to stay informed and aware and do the work on your terms, knowing it has to be done anyway, rather than having the circumstances dictate when you least expect them to.

GREEN LIGHT

From an energy and resource standpoint, doing your own repairs from materials at hand defers a contractor's trip to your house and your missed work time waiting to meet this contractor. You become more capable and have conserved your money. And if you have kids, you can pass these skills down to them.

Replacing a Light Fixture

- Be prepared for a few surprises when you tear open a wall or ceiling or snoop around an attic, especially in an older home that has been previously worked on.
- Code changes over the years will turn a simple job such as replacing a light fixture into a more complicated job.
- When some of your work involves concealed areas, figure on spending more time than you expected to; an assistant is always helpful, too.

Home Repairing Is a Learning Process

- Be flexible when approaching your repairs—your one trip to the store might turn into several.
- For some jobs, have a contingency plan in case water or heat has to stay off overnight or your painting is taking longer to dry than you expected.
- Know when to walk away from a frustrating repair—it really will look different after a meal or some sleep.
- If you're getting in too deep and need expertise, back out and hire some help.

SIZING UP THE PROBLEM

Determine which problems need your attention and which ones can wait for professional help

Whether you own a home or rent, problems arise: Drains and roofs leak, often at bad times (is there ever a good time?), the power goes out, the front door won't lock. A slow dripping faucet you can live with—forever, if you want—but rain pouring in through a hole in the roof from a now-detached tree branch can't wait at all. Solving a repair problem is no different than figuring out a math problem. First, read and assess the situation. Then try some quick solutions and reassess everything. Try the solutions you thought you could avoid, and if everything fails, do damage control to stabilize the problem until you have the tools, materials, or help to fix it.

You will run into new problems, but you can apply your

Categorizing a Problem

- Categorize each home-related problem to determine how and when you'll resolve it.
- A life-threatening or severe house problem calls for a more critical response than one that doesn't pose any danger.
- Some situations call for expertise beyond your own—don't hesitate to call a contractor, but understand contractors can be expensive.
- If you cannot completely solve the problem, try a temporary fix—such as shutting off water to a leaking pipe—until expert help arrives.

Basement Flooding

- If the flooding is due to extreme rain, block off any leaking windows and doors as best as possible and monitor the situation until the rain stops.
- Broken or backed-up sewer lines are unsanitary and unhealthy; call a plumber for repairs and for a clean-up contractor recommendation.
- If a plumbing pipe breaks, turn off the main water shutoff and call your plumber.
- Know where all your water shutoffs are located and confirm they're in good working order before you need them.

problem-solving skills to them and often come up with a reasonable solution. It might not be the fastest or most elegant, but that's not important. Speed and elegance will come later. For now, you took care of it, even if it meant calling someone else to do it. That's a perfectly legitimate solution if you've tried everything you could think of. And remember, online searching dramatically produces more solutions than any one of us can think of.

MAKE IT EASY

If the problem isn't endangering you or causing increasing damage—flooding water, for instance—and you're feeling time pressures to be elsewhere or get something else done, then walk away from it. Fatigue, hunger, and time concerns will negatively affect your actions and can make things worse. Some things you can always fix later.

Power Out

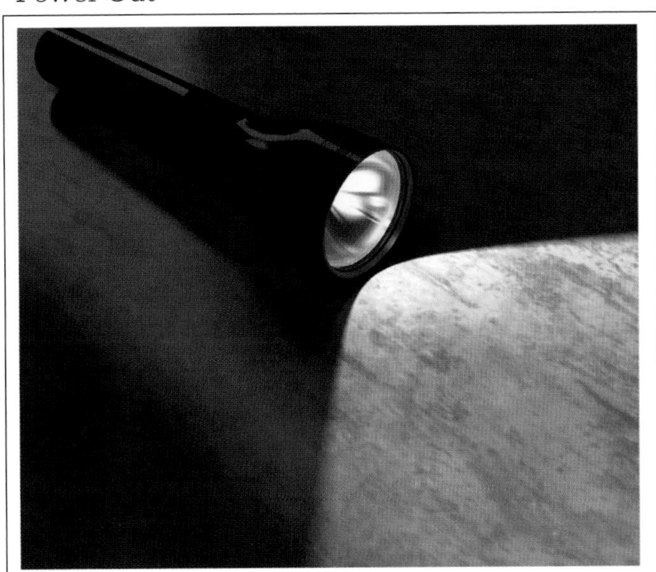

- If your neighbors are also without power, the outage isn't isolated to your house but can be weather-related or due to faulty utility equipment.
- If just your house has lost all its power, there is probably a problem with your main power lines.
- When only one area of power goes out, a fuse has burned out or a circuit breaker has tripped.
- When an overloaded circuit shuts down, figure out what you last turned on that caused the loss of power.

When to Report an Emergency

- Contractors are available for after-hours emergencies but at a steep price, so consider whether your problem really needs an immediate response.
- The time to find twenty-four-hour emergency help is when you don't need it—read online reports, contact the Better Business Bureau, and check that an advertised contractor's license and bond are current.
- Determine what you can hold off on fixing until normal business hours to offer a greater range of contractors to call—and better prices, too.

WATER LEAKS

Know which messes should get cleaned up early and which ones can wait for a professional

Water is great when we want clean water in pipes, dirty water in drain lines, and rain water outside. When a pipe or drain line leaks, however, and the rain comes through the ceiling, we panic. Water damages walls, floors, furniture, and anything else that gets in its way. Some leaks won't damage much if they occur in a burst pipe in a crawl space, but they will require all the water to the house to be shut off until they're repaired. In that sense, such leaks have a lot of impact, especially if one occurs in the middle of Thanksgiving dinner. Knowing some quick fixes will get you through until you or a plumber—and not a plumber on holiday rates—can do a permanent repair.

Leaking Pipe

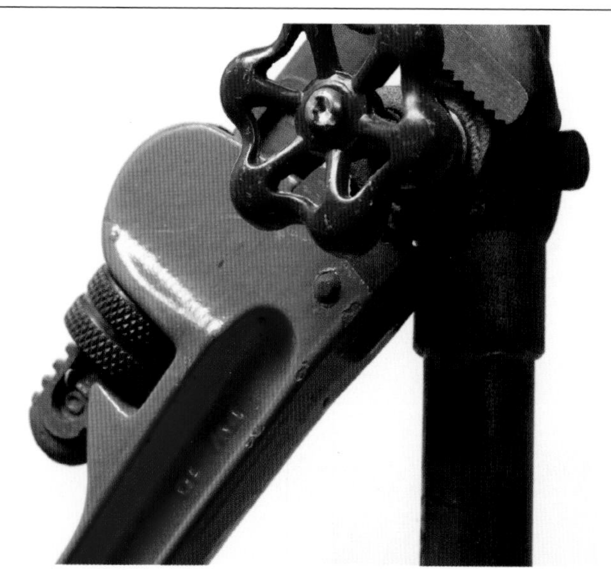

- Shut the water off at the fixture if the leak is out of a faucet.
- Shut the water to the house off if a pipe is leaking excessively.
- If the pipe is accessible, wrap it tightly with Wrap and Seal Emergency Tape, similar repair tape, or even a bicycle inner tube secured with wire.
- The leaking water can be caught in a bucket, which buys you time until the pipe is repaired.
- Most roof leaks aren't emergencies unless a tree branch crashes on it, leaving a large hole.

Leaking Roof

- In open attics, place a bucket under the leak and time the water going into it to determine how often it needs to be emptied.
- A minor, slow leak can even be monitored until repaired as long as the water doesn't damage any exposed wood on its way into a bucket.
- If the roof isn't accessible through an unfinished attic space, make your best guess as to where the leak originates and toss tarps or plastic over that area.

Toilets can leak, too, but if you have more than one, the leaking culprit can be shut off and addressed later.

Roof leaks are problematic and should be considered on a case-by-case basis. An unfinished attic space allows you to see where the water is leaking in; a bucket will hold you over until the roof can get patched. Failing that, a tarp on the roof will do the job. The point is to limit the water's damage if you can't stop the water itself. Taking interim steps is better than doing nothing at all.

YELLOW LIGHT

You don't want to discover you don't know where your main water shut-off valve is when you need it. Find it and test it yearly; it can be tight if it hasn't been used in some years. Spray it with a lubricant if it's sticking. Show other family members its location as well.

Leaking Toilets

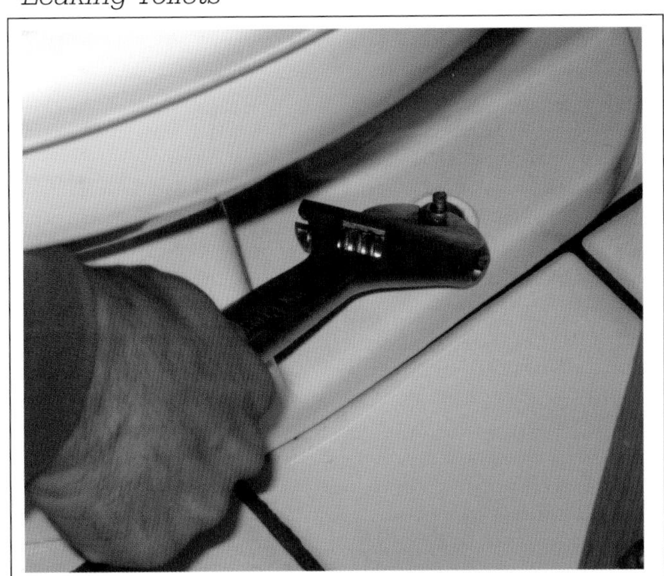

- To temporarily remedy a slow leaking toilet tank, catch the water with a bucket or plastic container.
- If the cold water supply line develops a small leak, place a plastic container under it as well, but address this soon—this valve is important, and you don't want it failing on you.
- When the leak is at the base of the bowl at the floor, shut the water off, flush the water out of the toilet, and stop using it until the toilet can be removed and reset.
- Running water inside the tank wastes water but can be fixed when time permits.

Preparing for Big Leaks

- Big leaks can be very damaging when they show up on floors and countertops instead of staying inside pipes and fixtures.
- Stop the flow of water as quickly as possible, which is at least as important as fixing the leak itself.
- You might have a leak and not even know it—check your water bill for any unexplained increases in water usage.
- Find a plumber now so you're not looking when you do need a plumber.

GAS LEAKS

Any natural gas smell is dangerous and must be taken seriously and treated immediately

The majority of homes are heated with natural gas. In its natural state, it's odorless and colorless. If it was leaking, how would anyone know? The stink made by natural gas is from an additive called mercaptan. A small amount goes a long way, and it's a good thing. A gas leak in the vicinity of an active pilot light in a water heater or stove can have explosive consequences.

It's difficult to lay down any hard and fast rules regarding how strong of a gas smell should drive you out of the house. If a pilot light goes out, and you detect the faint odor of gas, it's probably from the pilot and nothing else.

A gas pipe doesn't suddenly rupture in your house unless it's somehow damaged by a nearby activity. It's critical that

Rules for Gas Leaks

- Whenever you smell a gas leak, turn the gas to that particular appliance off immediately and air the area out.
- A gas smell might be only a pilot light that's gone out, which can be relit after the room has aired out.
- If the leak is much more noticeable, shut off the gas to the house immediately, and call the gas company.
- Never try to repair a leaky gas line with any kind of temporary patch.

Gas Furnaces

- Gas furnaces are normally safe and do not present fire hazards, but there are exceptions.
- Some furnace models have been subject to manufacturer recalls—check yours online or talk with your heating company if you're uncertain about it.
- A more probable issue with gas furnaces is carbon monoxide poisoning from incomplete gas combustion.
- Regular furnace servicing should preempt any carbon monoxide problems.

you know how to shut the gas off at the meter and at each gas appliance!

An overwhelming gas smell sends a clear message: Get out and call the gas company from a neighbor's house or a cell phone. Don't go back inside. If the meter is outside and you can shut the gas off, do so. Otherwise, let the gas company do it. It would prefer that as well.

GREEN LIGHT

The next time you have your gas furnace serviced, ask the repair technician to show you how to shut the gas off at the meter if you're uncomfortable doing it yourself. Discuss what emergency measures to take and listen to the recommendations. The gas company wants you to stay safe, too.

Gas Appliances

- Read your appliance owner's manual for proper operation and maintenance.
- Aside from regular gas furnace servicing, check gas appliances as well for clean combustion and operation.
- Think twice before installing a ventless gas fireplace; it passes many building codes but poses some risk factors not present with vented systems.
- Pilot lights can ignite solvent fumes—turn them off when volatile solvent fumes are present (during floor refinishing, for example).

Understanding Your Gas Meter

- Your gas shut-off valve is normally located near the gas meter. Find yours or ask your appliance technician.
- Keep a 12-inch adjustable wrench close by to shut off the gas in case of an emergency or suspected leak.
- FEMA suggests testing your gas shut-off valve periodically (moving the shut-off valve barely 1/8th of a turn—turning it farther will shut off the pilot lights) to confirm it moves easily.
- Have a qualified gas technician check for leaks, turn the gas back on, and relight the pilot lights should you ever have to turn the gas off completely.

STORM DAMAGE

Storms and bad weather have plans of their own, so make a game plan for repairs

Weather can be the biggest culprit in causing damage to your home. A harsh freeze can burst a pipe. Ice-laden tree branches can snap off and hit your roof or block a driveway. If enough wind shows up, there go the phone wires.

One of your best recourses when confronted with severe weather is to be prepared for it, but few people are.

A main concern for homeowners is having live power lines go down near their homes. Current can carry across the ground, particularly if the ground is wet, and can shock or electrocute anyone standing nearby. Watch out for power lines before tangling with tree branches. Your power company will find them eventually, but try and report them any-

Tree Damage

- A fallen tree or branch can render a house unlivable and even unsafe for occupation until the damage is evaluated and repaired.
- If you cut away part of the damaged tree, be careful, as this could cause more damage to your home by the remaining section of the tree.
- The best way to avoid tree damage is to keep your trees healthy and remove ones that appear problematic; call in a tree service firm if needed.

Securing Roof Damage

- Clear the damaged area and cover with a tarp as soon as possible.
- If there's a gaping hole in the roof, nail a tarp over the hole.
- Tuck the tarp under a row of shingles above the opening so water will flow over it and not under it.
- Pull the tarp so it's taut and doesn't allow water to pool in the center.

way, especially if they're sparking and endangering you or your neighbors.

What happens if your home is damaged by extreme weather? Act sensibly and evaluate the problem. Assess the risks of walking on a roof to remove debris yourself. If the main service line that supplies your house with water bursts, shut the water off at the meter, call a plumber, and find a hotel room since you're out of water until the pipe is replaced.

MAKE IT EASY

Review your insurance policy for weather-related damage and update it if you're living in an area susceptible to a particular weather problem. Weigh the costs against the chance of ever needing the coverage. In the event of damage, know who to call and follow the claims procedures to the letter. And remember to photograph all damages for the insurance company.

Power Line Dangers

- Stay away from downed electrical lines as a current can spread across wet ground.
- Even if the power is shut off and the wires are dead, you can be fined for working around high voltage wiring.
- Don't be tempted to move a downed power line even if you're using nonconductive materials such as wood—if they have any moisture on them, they can conduct the current.
- Don't drive over power lines—they can still be energized and can get entangled with your car.

Broken Window Glass

- If you can't get outside to cover a broken window, tape plastic all around the inside as tightly as you can and place a towel at the bottom of the window.
- Exterior plywood will withstand wet weather longer than any interior sheathing, especially the composite products such as particle board, but for short term, any wood product will do.
- Don't use sheets of drywall—they'll deteriorate in the rain.
- Carefully remove any remaining loose glass shards from the window.

SECURING A DAMAGED HOUSE

Secure broken or damaged windows and doors quickly to ensure your safety

Property gets damaged all the time, sometimes by animals, sometimes by weather, sometimes by other people. A kicked-in door or broken window makes for easy entrance into the house and can leave you vulnerable, so you must secure your home until you can render a complete repair or replacement.

Damaged doors and windows can be secured without marring up the surrounding woodwork too much. Boards can be fastened across them to substitute for broken locks, and plywood can take the place of broken glass. If a door is so badly damaged it can't be closed, full sheets of plywood can secure an opening.

Securing Your Windows

- A damaged sliding wood window or out-swinging casement can be secured with blocks of wood screwed to the sash and the jamb or lengths of wood wedged in the tracks to prevent the sash from moving.
- With sliding metal windows, install sheet metal screws in the track to prevent sliding.
- Block sliding vinyl windows with full lengths of wood and avoid screwing anything into any part of the window.

Quick Door Lock Repairs

- Clean out the damaged wood from the jamb; replace the jamb door strike screws with long drywall screws, placing a wood shim behind the strike so it will line up with the door.
- Straighten out and reinforce the lock with the longer drywall screws.
- Close the door and screw a 12-inch block of wood into the wall stud on the lock side so the wood crosses the door.
- Screw the block into the door, noting that this door cannot be used for an emergency exit until it's replaced or repaired.

If a garage door has been damaged or won't close, there is often a simple solution—adjusting the sensors, for instance—but if not, modern door openers have release mechanisms for manual operation. Knowing how your door opener works ahead of time will save you some headaches later when it doesn't.

MAKE IT EASY

Check your homeowner's policy for damage coverage. Your policy should cover the cost of door replacement plus any damage to the jamb, trim, and lock and the follow-up painting, but confirm with your agent first.

Garage Door Opener Manual Override

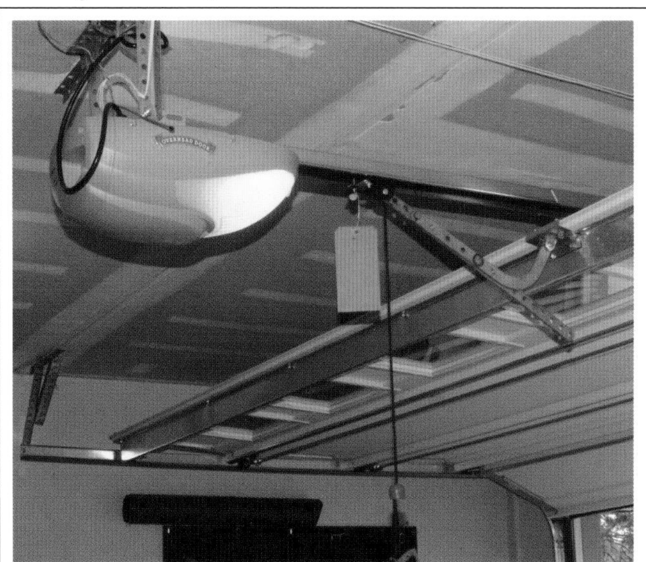

- Automatic garage door openers won't function if the power goes out.
- Pull the release cord or lever on the track near the top of the garage door to release it from the automatic opener and allow you to manually open and close the door until it can be repaired.
- Pulling this release again—follow the manufacturer's instructions—will return the door to machine control and keep it closed and reasonably secure.

Home Security Systems

- According to the U.S. Department of Justice, as many as 98 percent of all alarm calls involving security systems are false alarms costing millions of dollars for police to respond.
- As a result of so many false alarms, some cities are fining homeowners and alarm companies, and others are not responding to residential alarms.
- An alarm system's effectiveness from a burglary standpoint is dubious, but helpful for fire prevention since a detector sensing smoke followed by notification of the fire department can prevent major fire damage.

TOOL ERGONOMICS & SAFETY

Learn the right way to choose and use tools to make your job safe and easy

A good set of comfortable tools used safely can solve all kinds of repair problems. A tool that doesn't fit your grip is not only uncomfortable, but can also be unsafe if it slips during use. It should fit you as close to your dimensions as possible. Given the choice of sizes, weights, and grips available, suitable tools are easy to find.

Size adaptations are also possible with existing tools. For instance, tools can be adapted to large hands by wrapping them with foam handlebar from a bike shop or by wearing work gloves. Those with small hands will have to test out different brands to find the best fit.

A tool's weight can help it do its job or might mean it's

What Is an Ergonomic Tool?

- Ergonomically designed tools help avoid extreme gripping force, awkward hand positions, excessive vibrations, and compression of nerves and blood vessels.
- A textured rubber handle allows for a good grip without applying a lot of pressure, and there is less chance of the tool slipping.
- The ideal ergonomic power tool is light enough to use with one hand, but this isn't practical for some larger, heavier tools.
- An ergonomic tool that helps avoid fatigue is more efficient.

Tool with a Proper Grip

- For select tools use a power grip—fingers wrapped around the tool and the thumb against it or wrapped around resting on the fingers—for most of your heavy work.
- Short-handled tools may be less stressful than long-handled tools.
- Tools made with compressible grip wrapping are usually easier on the hand than hard plastic.
- The tool should grip comfortably in either hand while avoiding bending or rotating the wrist.

built for frequent, commercial use. A "heavy duty" commercial tool lasts longer than a less expensive version, and one that's too heavy for someone in average physical condition causes strain and should be avoided. Most hand tools, with the exception of some hammers, are light enough for comfortable use.

A damaged or worn tool increases the strain on you when using it and should either be repaired or discarded. Most quality hand tools will last for years with little maintenance.

Weight of a Tool

- Proper swinging technique allows the weight of a striking tool (hammers, axes) to do a large amount of the work.
- Some tools (sledge hammers, for instance) must to be heavy to do their jobs.
- Overhead work with heavy tools is the most strenuous, so to avoid accidents, take breaks when fatigued.
- If a tool, such as a gas-powered chainsaw, is too heavy to safely handle, find a smaller, lighter version, even if the job takes longer to do.

YELLOW LIGHT

A true ergonomic tool design—where the tool fits your hand rather than your hand adapting to the tool—allows for the tool usage and function, not just the grip. Short-handled pliers, for example, will be difficult to use regardless of added padding.

Maintaining a Hand Tool

- Be sure all tool handles are secure, blades sharp, and all moving parts shift freely.
- Store tools in a dry location; to avoid corrosion, spray the tool with WD-40 periodically.
- If a tool is beyond repair, discard it (a damaged tool is a liability).
- A wire brush attachment on a bench grinder will clean up the metal parts on any hand tool, removing corrosion and dirt and bringing back a newer appearance.

HAND TOOLS

Collect the must-have tools you need for most small jobs around the house

Hand tools are a bargain—what else last as long yet cost so little? Each job dictates the needed tools, but a core set will fix most problems. A well-stocked toolbox or tool bucket needs basic tools that pound, cut, drive and loosen fasteners, measure, grab, pull, and scrape. Tools are extensions of our hands and offer clout in the repair world we would never have otherwise.

Must-have tool lists vary slightly from one another, but most include a hammer, screwdriver, pliers, measuring tape, an adjustable wrench, metal snips, putty knife, level, utility knife, paint scraper, nail set, and a simple pencil and pad of paper.

As your job list grows, you'll add to your tools, but more isn't necessarily better nor are multiple copies of the same

Hammers

- Heavier hammers drive nails faster than lighter hammers—use them for framing and other rough work.
- There is a hammer for every purpose, but a curved claw hammer will do most household jobs.
- You might find wood handles more comfortable than steel or fiberglass (if tool vibrations bother you, search online for "ergonomic hammers").
- To avoid accidents, never use a hammer with a loose or damaged head or a cracked handle.

Screwdrivers

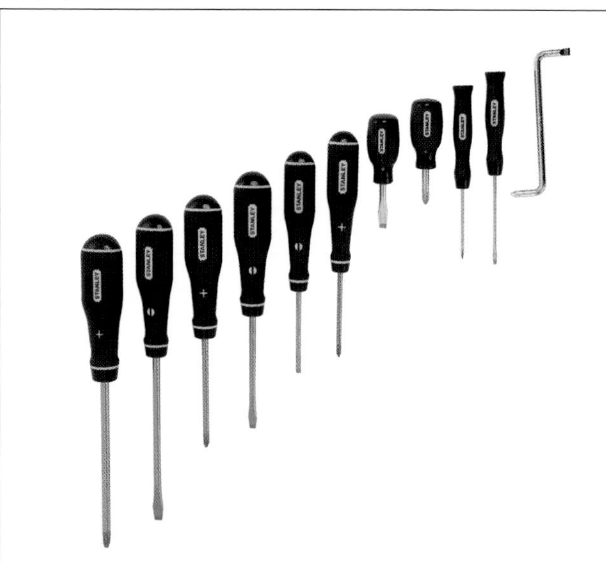

- Screwdrivers fit a number of different screw types, but the most common are slotted (regular) screws and Phillips screws.
- Combination or multi-bit screwdrivers, with both Phillips and slotted drivers, are handy tools; some versions offer other interchangeable bits, which can also be used in electric drills, in a carrying case.
- The tips of cheap screwdrivers will quickly become bent and useless.
- Consider a screwdriver with a magnetized tip for holding steel screws.

tool. You can spend as little as a dollar per tool or much more, but either way you get what you pay for. A good hammer will last a lifetime, but a cheap hammer will bend or break, a dangerous possibility while pounding. Although tools are available to buy online, there's no replacement for handling a tool and getting a feel for it in your hand to be sure it fits for you.

MAKE IT EASY

Combination hand tool kits group frequently used tools in one convenient package and make a great extra tool set for a kitchen or storage closet. These are moderate quality and are not replacements for higher quality individual tools. Use them for small jobs—tightening a screw, removing a picture hook, etc.

Pliers

- Pliers grip, bend, and sometimes cut, but too often they're used to tighten nuts and bolts, normally a job for a wrench.
- A toolbox should have needle-nose pliers for doing precision work and handling small objects, channel locks for plumbing repairs, and a pair of lineman's pliers for both cutting and gripping.
- A curved needle-nose pliers has a bent end for reaching around obstacles.
- Vise-type locking pliers lock in place while gripping, leaving both hands free.

Wrenches

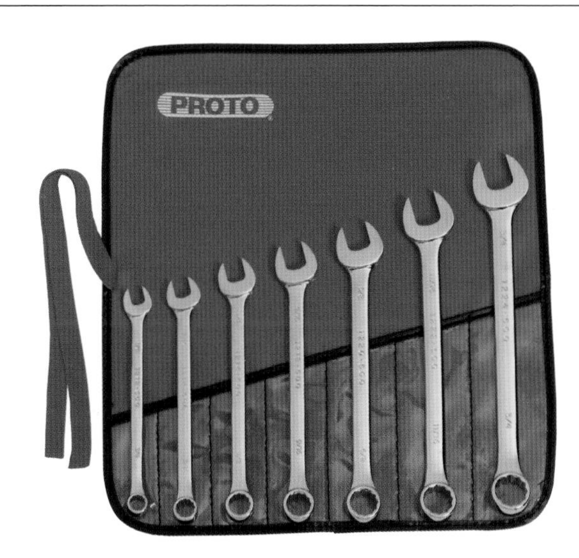

- Fixed wrenches, with an opening on each end, will fit only either inch-based (S.A.E.) or metric-sized fasteners—they are not interchangeable even if they look close in size.
- An adjustable wrench, whose head adjusts to accommodate different sizes of fasteners of any type, can take the place of multiple individual wrenches.
- A hex head wrench, or Allen wrench, adjusts hex-shaped bolts, often found on "some assembly required" furniture.
- Match the wrench to the fastener for a tight, secure fit.

TOOLS TO CUT, SCRAPE, PRY

Know what essential sharp, tough tools are needed for cutting a job down to size

When cutting through material, the sharper the tool the better for an easier, more precise, and safer job. Dull blades increase a job's difficulty and time needed to finish and are more apt to slip and injure you. Blades are made to be sharpened, although most are eventually replaced. It takes some practice to get a good edge on a scraper or some knife blades using a file or sharpening stone, but sharpening when a blade begins going dull increases the life of these tools and is a worthwhile skill.

Handsaws can rip through framing lumber or make the most delicate cuts in wood trim and molding. Although power saws are used for most cutting now, there are times when only a

Cutting Tools

- Quality tools use harder steel and maintain their sharp edges longer than less expensive cutting tools.
- A retractable utility knife, which comes with extra blades in its handle, strips wire and cuts cardboard, wallboard, carpet, and some plastics—it's a must for your toolbox.
- Tin snips or tinner's shears shape sheet metal and can cut rope, cardboard, metal weather stripping, and small wires.
- Longer-handled tin snips offer more leverage but might be less comfortable for smaller hands.

Pry Bars

- You should need only one pry bar of any given size as they last forever and almost never break.
- Some very flat and thin pry bars can also be used as scrapers.
- A pry bar that fits in a toolbox is as big as you need for most jobs.
- Woodwork always has more nails than you might expect—pry carefully to avoid splitting the wood.

handsaw will do. Besides, why drag out a power miter saw when a handsaw can make quick work of a single cut?

Pry bars range from mild-mannered to aggressive. Some remove finish nails, and others pull up entire floors. Most home use calls for flat pry bars, including a 5- to 6-inch-long mini bar. Flat bars are thin enough to slip behind wood trim and baseboards and remove them with limited damage.

Scrapers are highly versatile. They can remove paint and floor finishes and shave down wood. A 1 1/2-inch-wide hook scraper is a traditional choice.

ZOOM

Pry bars are all about leverage. A long bar has more than a short bar, but not every job needs a longer bar. Use a block of wood under whichever end of the pry bar presses against a finished surface to avoid damage. Pry slowly and steadily.

Saws and Files

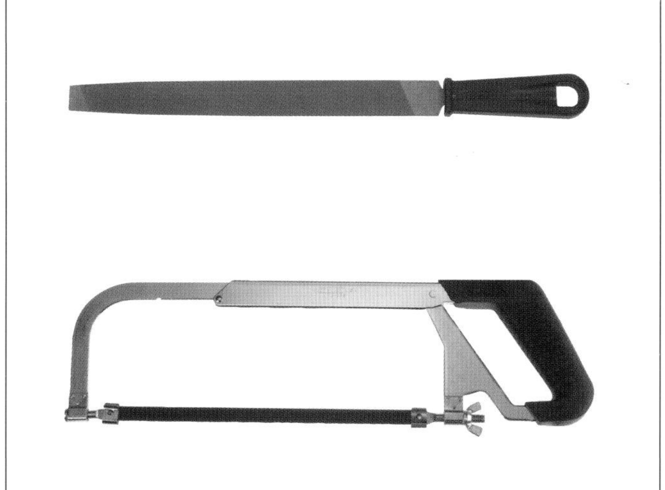

- It's worth spending a little extra on a carpenter-quality handsaw, which will keep sharp cutting teeth much longer than cheaper saws.
- Crosscut saws have teeth that cut across wood grain, ripsaws cut with the grain, and a bow saw is used for landscape trimming, although a crosscut saw can cut through branches.
- Japanese saws have thin blades and are ideal for precise cutting.
- Hack saws, normally used on metal, also offer precision cutting on thin sections of wood.

Scrapers

- It's a good idea to buy extra blades when buying a scraper so you always have sharp replacements on hand.
- The simplest scraper is a rectangular cabinet scraper, used for scraping wood smooth prior to finishing; it also removes varnish.
- A 1 1/2-inch-wide hook scraper is simple, versatile, and easy to handle while scraper kits offer multiple, changeable blades for different shapes and cuts of wood molding.
- Carbide scrapers have the longest-lasting blades.

BASIC SMALL TOOLS

These seemingly insignificant tools play a big role in many home repair jobs

Many repair jobs require measuring. You need to know the size of a room before buying paint or the dimensions of a window to order a blind. A number of tools are available for measuring, but a tape measure is the simplest and most compact. A longer tape measure is more versatile than a short one—look for a 25-foot metal retractable-tape model with a locking mechanism. Most longer tape measures display measurements down to 1/16 inch, more than enough precision for home repairs.

Chisels are used for removing small sections of wood, stone, concrete, or mortar. Practice using a wood chisel on a piece of scrap wood first. Stone chisels are heavier and less refined-

Tape Measure

- It's worth having one or two small tape measures around the house for impromptu measuring jobs in addition to a 25-foot model.
- Take your time reading your measurements—you don't want to order window blinds based on the wrong dimensions.
- Laser measurers calculate distance, square footage, and volume, depending on the model.
- Wipe the metal tape with a clean cloth, lightly coat with spray silicone, don't allow the tape to retract too quickly, and replace a tape measure that has cracked tape.

Heat Gun

- Heat guns produce extremely hot air of varying temperatures for removing paint, drying out wet wood, speeding up paint drying, softening adhesives, and thawing frozen pipes.
- Heat guns are safer than open flame torches but still require caution.
- The materials being heated can produce disagreeable if not toxic fumes, requiring you to wear an appropriate respirator.
- Lead-based paint will vaporize at temperatures over 1,100 degrees F—use a heat gun with a lower setting for this job.

looking than wood chisels, but there's no other hand tool as effective for attacking concrete cracks that need widening or cleaning out.

Nail sets are metal punches used to force finish nails below the surface of wood. This allows the space above the head to be filled for a smooth finish. Individual nail sets are available, but buy a set of three (1/32-, 2/32-, and 3/32-inch).

A basic circuit tester indicates if electricity is present at a receptacle or a device. It's an inexpensive but critical safety device that every electrician uses and depends on.

MAKE IT EASY

Use a large framing nail as a nail set. File the sharp end down a bit first. This is not as sturdy as a normal nail set, but it will do in a pinch. Using a screwdriver as a chisel, however, is a great way to ruin the screwdriver.

Chisels

- You're better off buying one or two good wood chisels than a set of poor quality tools.
- Using a wood chisel is an art requiring practice and patience—try it on a piece of scrap wood first until you're comfortable with it.
- Inexpensive chisels are not easy to maintain, get nicked easily, and eventually become useless for precision work, so keep in mind value when assessing cost.
- Protect the sharp ends of chisels when stored so they don't get dull.

Nail Sets

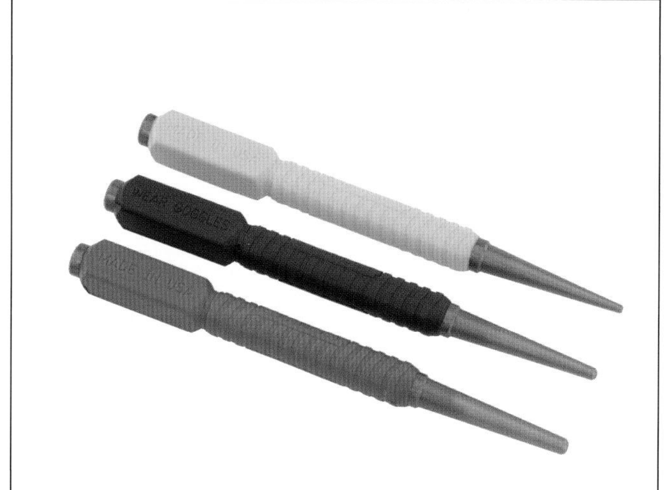

- If you buy a kit of three different-size nail sets, you shouldn't have any trouble matching a nail with the proper-size set.
- Don't treat a nail set as a chisel to chip away at stone or concrete—this will ruin the finish end.
- Use a nail set to make a starter or pilot hole for large screws if a drill isn't handy.
- Nail sets can punch holes in sheet metal and leather as well as wood.

POWER TOOLS

These tools perform with a speed and accuracy you can't match by hand

Imagine mixing a smoothie by hand. You can do it, but will you match the consistency and texture of a blender in the same amount of time? The same holds true for power tools. They will outcut, outdrill, and outsand their hand counterparts every time. There is no reason not to use them and no virtue in avoiding them, although they can be more bother than they're worth for very small jobs, especially if it means dragging out extension cords.

Some power tools do multiple jobs. A drill makes holes, but it also buffs, sands, and grinds with the right attachments. Power tools are loud, so ear protection is a must, even for quick jobs. They are also dangerous. It's one thing to slip

Drills

- Drills perform multiple tasks such as making holes, driving screws, and running buffing pads and sandpaper discs, but you're better off using a buffer/polisher or sander for these two tasks.
- A medium-duty, 1/2-inch variable-speed drill will handle most household jobs.
- Cordless drills are convenient, but require regular battery recharging.
- The material you drill through determines the type of bit you'll need—one type does not work for all jobs.

Sanders

- Sanders throw out a lot of dust, but make quick work out of tedious jobs.
- A belt sander removes material quickly, while a random-orbit disc sander is less aggressive, leaving a smoother finish. Use palm sanders for finish work.
- Dust bags will not collect all the dust produced by a power sander, so be prepared for some additional clean-up with a vacuum cleaner or broom.
- Practice sanding scrap wood, applying light pressure until you're comfortable with the tool.

while using a handsaw, quite another to slip with an electric circular saw running, probably the most dangerous power tool commonly used. Caution and care must accompany power tool use, but the advantages they bring more than outweigh the extra vigilance on your part.

Power tools create dust like hand tools do, but it spreads out over a larger area. There is more clean-up, but a vacuum cleaner solves that problem faster than a broom and dust pan.

YELLOW LIGHT

Sanders are loud, so wear ear protection as well as a dust mask. Do not wear work gloves with either belt or disc sanders—they can get caught in the moving parts and pull your fingers into the tool. Sandals are for the beach; wear good work shoes around power tools.

Saws

- An electric saw is as dangerous as it is useful—always treat it with respect and care.
- Use circular saws to cut lumber and plywood or a modestly priced power miter saw (an excellent choice for nonprofessionals) to safely cut framing lumber and finish trim.
- Match the saw blade to the type of cutting you're doing—plywood, lumber, finish work, or metal.
- Don't force the saw; move it as it only makes its cut.

Vacuum Cleaners and Blowers

- Inexpensive canister vacuum cleaners, which put out less exhaust dust and are more portable, are convenient and durable ways to clean up a mess
- If your vacuum uses disposable bags, buy extra ones when you purchase the vacuum cleaner.
- HEPA vacuums trap finer dust and allergen particles than standard vacuum cleaners.
- Electric blowers are good tools for blowing dust out of standard household vacuum cleaners, especially the hoses.

TOOLS YOU RENT

Some tools are impractical to own but perfect to rent for the occasional use

There is little point in owning a tool you rarely need or use only once. A good rental shop offers a range of tools at affordable prices along with the additional necessities: extension cords, sandpaper, drill bits, saw blades, and so on.

If possible, reserve the tool in advance. You and most other homeowner customers will be out in force on the weekends and all looking for the same tools. Commercial users do most of their renting during the working weekdays.

Rental shops vary from the single-owner shops to chain stores. A good shop has a policy of checking all tools upon return and before setting them out for another renter. Electrical cords should be intact and not taped or patched, and

Wallpaper Steamer

- Wallpaper steamers are simple and foolproof: Add water, heat up, and steam off multiple layers of wallpaper, being careful to cover the floor area completely with plastic and drop cloths.
- Follow the manufacturer's instructions for adding water—the unit is extremely hot, and you can be exposed to scalding water.
- Rental units are larger and more robust than many consumer models.
- Use to remove painted non-asbestos "popcorn" ceilings for a dust-free job.

Chainsaws

- A gas chainsaw is a good tool to rent for major tree work and is also a heavy tool that must be used carefully.
- Electric chainsaws are lighter, safer than, and not as powerful as gas-powered models but are inexpensive enough to consider purchasing for regular pruning.
- Never use a chainsaw when you're tired—it's too dangerous.
- Chainsaws can also cut through wood posts and beams, provided they're free of nails and other metal objects.

the rental clerk should give you a run-through on each tool's operation and explain your liability (be sure to read the rental contract so you understand your obligations).

Depending on the tool and the length of the rental period, you might be better off purchasing some tools. Compare both costs before deciding either way. For shops closed on Sundays, some rentals are available at reduced rates if you pick up the tool on a Saturday at closing time and return it first thing Monday morning.

YELLOW LIGHT

If you have problems with your rented tool (e.g., the motor slows down or is otherwise malfunctioning), stop immediately and call the shop. These tools are heavily used and even with inspection can break down unexpectedly. If you're not at fault, the shop should credit you for lost time and replace the tool.

Pressure Washer

- A pressure washer is rated by pressure or PSI (pounds per square inch) and the volume of water—GPM or gallons per minute—that flows through it.
- Pressure that's too low won't clean well and too high can damage some surfaces (1,800–2,600 PSI will do most jobs).
- When washing wood surfaces, start from 2 or 3 feet away and gradually move forward to test the pressure.
- Pressure washing will not remove all dirt and grime—additional scrubbing might be required.

Nail Guns

- Nail guns allow house framers to avoid hand nailing and the fatigue that comes with it.
- Air-powered nail guns, which require an air compressor, make short work of big nailing jobs; cordless gas-powered guns are available for finish work and framing work, but they're expensive and worth renting.
- For strictly light finish nailing, an electric nail gun for under $100 is worth purchasing.
- Nail guns nails are sold by the box—buy extra just in case, as they are always returnable.

SAFETY GEAR

Doing a job right means taking the precautions to protect yourself, too

Using tools means dust, noise, nicks, scrapes, and flying debris with the user on the receiving end. It's easy to discount safety, but for a few minutes and a few dollars, you can protect your sight, hearing, skin, and lungs.

Both power tools and excessive hammering assault our hearing. Once it's damaged, it doesn't come back. Even occasional power tool usage calls for hearing protection. Consider this: Normal conversation registers at 60 decibels, a power saw—probably the worst power tool offender—registers at over 100. Ear protection is simple and inexpensive and quickly becomes second nature when doing house projects. Protective equipment ranges from disposable ear plugs to state-of-the-art ear

Ear Protection

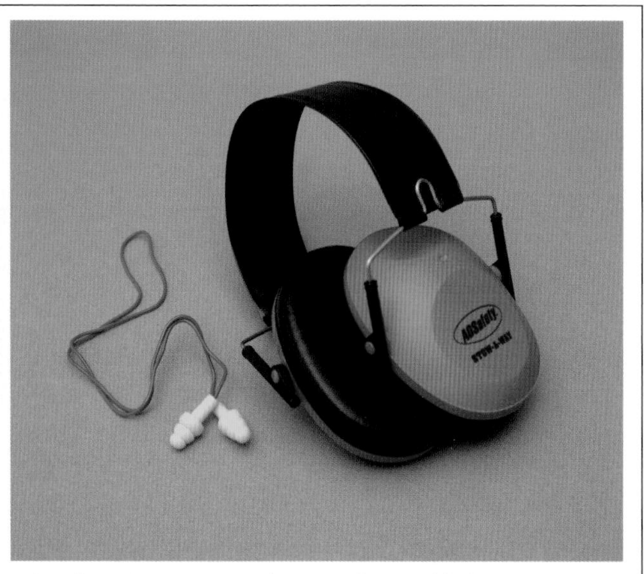

- Loudness is represented by decibels—the higher the decibels, the harder it is on your hearing—whether the noise source is from power tools or excessive hammering.
- Ear muff protectors are the most comfortable form of hearing protection.
- Disposable foam ear plugs can be washed and reused—just soak them in warm, soapy water and keep extra pairs on hand.
- Always keep a pair of disposable plugs with you—you never know when you'll run into prolonged noise exposure.

Eye Protection

- Safety glasses are available in both clear and tinted lenses for working outdoors in direct sunlight.
- Hammering, grinding, sawing, sanding, and power washing all have eye-damaging potential.
- Wear safety glasses when working with chemicals overhead, such as paint remover.
- Store safety glasses in an old sock to prevent scratching the lenses, and always replace cracked safety glasses.

muff–style devices with built-in AM/FM radios.

Vision is priceless, while eye protection costs next to nothing. Inexpensive safety glasses that also fit over regular eye glasses should be worn when chipping, chiseling, sawing, or doing any overhead work that blows dust or debris into your eyes.

Dust, smoke, and chemical vapors can also affect your health. Dust masks and appropriate respirators are a must when sanding or handling noxious solvents and paints.

Even tough hands need protection. Gloves are a bargain compared with damaged hands.

YELLOW LIGHT

Be sure to match the level of protection to the hazard produced by your work. Overkill wastes money, and too little protection doesn't keep you healthy. This is especially true when picking out the right respiratory protection.

Respiratory Protection

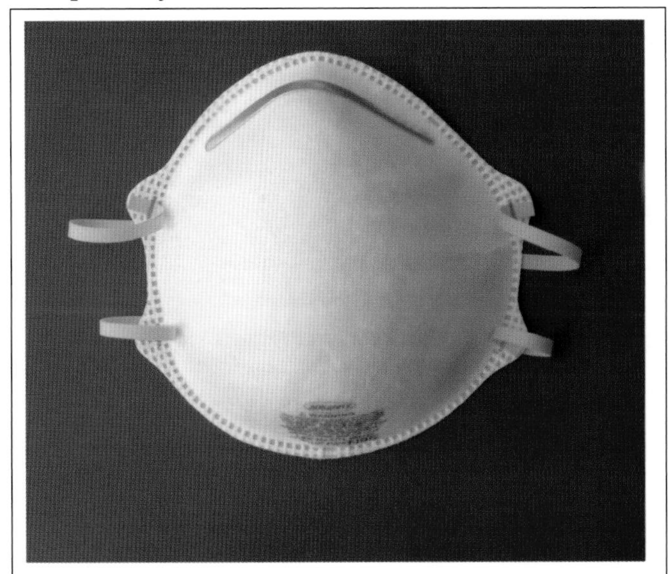

- Respiratory protection is available in disposable masks and full- and half-face rubber/plastic reusable respirators that use disposable filters.
- Respirators are not particularly comfortable, although the disposable style is more tolerable to wear.
- Buy dust masks (3M 8200 series is recommended) by the box so you don't run short; avoid light, paper dust masks, which provide little protection.
- Protection from smoke calls for HEPA filters and vapors for charcoal-activated filters.

Hand Protection

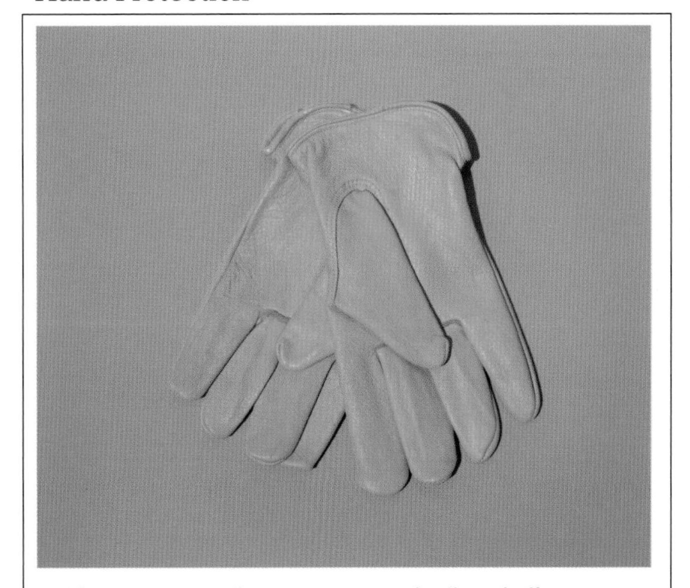

- Gloves wear out often—keep an extra pair or two around as replacements.
- For painting, wear lightweight cotton neoprene and PVC gloves when working with paint stripper, solvents, or washing solutions.
- Leather gloves protect against rough lumber and nails and offer some shock absorption when hammering.
- Never wear work gloves when using power tools with spinning blades, shafts, or similar moving parts—the gloves can get caught and can pull your hands into the tool.

SCREWS & BOLTS

An ancient and great invention, screws of every type and size keep it all together

Screws and bolts are necessities for putting projects together. A screw is a threaded fastener. Threads give a screw both penetrating and holding power. When screwing into wood, most screws require a pilot hole to be drilled first; otherwise, you risk splitting the wood, jamming the screw and often stripping its head. Some screws, notably drywall and sheet metal screws, are self-tapping, which means they do not require a pilot hole before being inserted and driven in. One advantage of screws over nails is that they can be removed and reinserted any number of times without damaging whatever the screw is fastening.

Screws vary by thickness, length, finish, and head type (slot-

Wood Screws

- Drill a pilot hole or starter hole—narrower than the shaft of the screw—to avoid splitting wood with a screw.
- Screw threads form an inclined plane that, when inserted into a piece of wood, forms a corresponding plane that holds the screw in place more tightly than a nail.
- Rubbing the screw threads on a bar of soap can make a screw easier to turn, while overtightening can snap its head off.
- Phillips head screws are easier to install and remove than slotted.

Drywall Screws

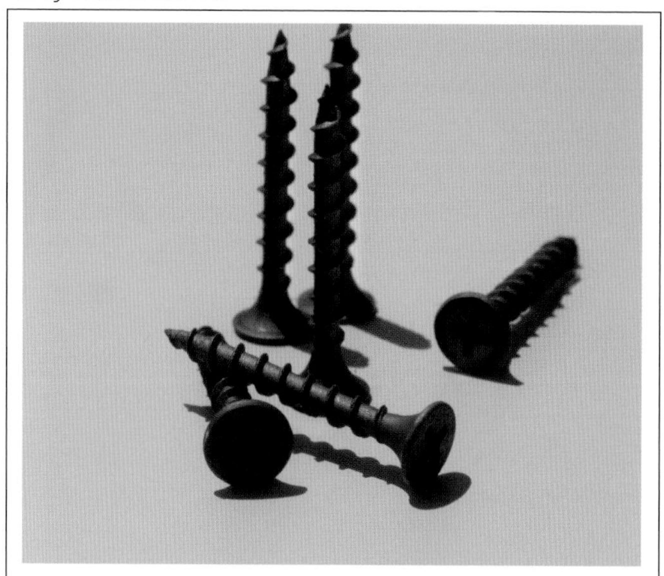

- Drywall screws are the universal fastener for nonmetals; case-hardened and designed for securing drywall, they're inexpensive, self-tapping in most materials, and screw in fast.
- These screws are sold by the pound; buy several sizes for various jobs.
- Do not use them for structural applications or outside—even the coated drywall screws will rust.
- They are available in lengths from 1 inch to 8 inches in both fine and coarse threads.

ted and Phillips being the most common) and are sold by the box, bag, or the piece. Flat-head screws lie flush, while oval and round-head screws protrude from the surface. Overwhelmingly, screws are made from some type of steel. Brass screws are used to attach visible hardware such as cabinet hinges and drawer pulls.

A bolt resembles a screw except the threads are narrower and do not cut through the material they're attached to. A bolt normally requires a threaded nut at the end of it to secure it. Bolts and nuts are sold in similar quantities as screws.

RED LIGHT

Drill a pilot hole too small, and you won't get the screw all the way in. If you get too much resistance, back the screw out and drill a slightly larger hole. Overtightening a screw can snap off its head. Be especially careful when tightening a screw with a drill.

Sheet Metal Screws

- Sheet metal screws are self-tapping—they cut their way through metal, plastic, and fiberglass—and are harder than wood screws.
- Small, flat-head sheet metal screws can be used to install metal weather stripping when nailing isn't practical.
- Use these screws to reattach downspouts to gutters, and when replacing corroded furnace or water heater venting; attach old sections to new with sheet metal screws.
- You can use these screws in wood but should not use a wood screw in sheet metal.

Nuts and Bolts

- Bolts and nuts come in standard (inches) and metric sizes—don't mix them together!
- To replace a bolt and nut, take both old ones to your hardware store and match each with its new counterpart.
- If either the bolt or nut threads are stripped—meaning they're worn smooth and no longer tighten—replace both bolt and nut.
- To keep track of bolts/nuts/washers during a disassembly, loosely connect them together so they stay in a grouping.

NAILS

For fastening wood to wood, nails are quick, fast, and cheap

Nails range in size and shape from the tiniest finish brads to 12-inch spiral spikes. They have round heads for when their appearance isn't important or distracting, and they have no head at all other than a dimpled end for finish work. These nails are driven below the surface of the wood so they won't be seen. Nails are manufactured for all purposes and applications, from withstanding exterior weather exposure to assembling wooden fruit boxes. It's best to use specific types of nails for specific jobs, but in some instances, you can substitute, for example, a longer nail for a shorter one. For other jobs—framing comes to mind—you must use the size and type of nail determined by your local building code.

Pay attention to a nail's size. Too long, and you can drive it all the way through a board, leaving the sharp end sticking

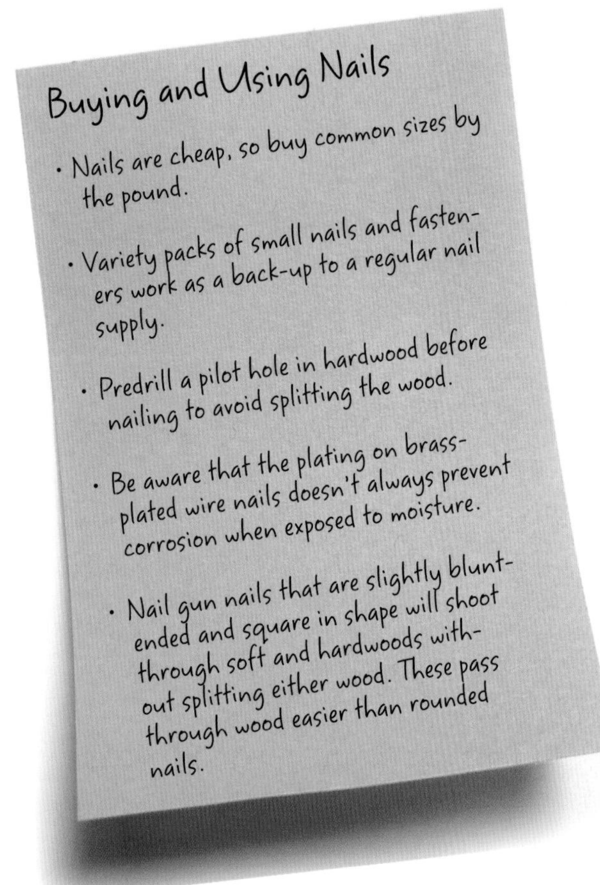

Framing Nails

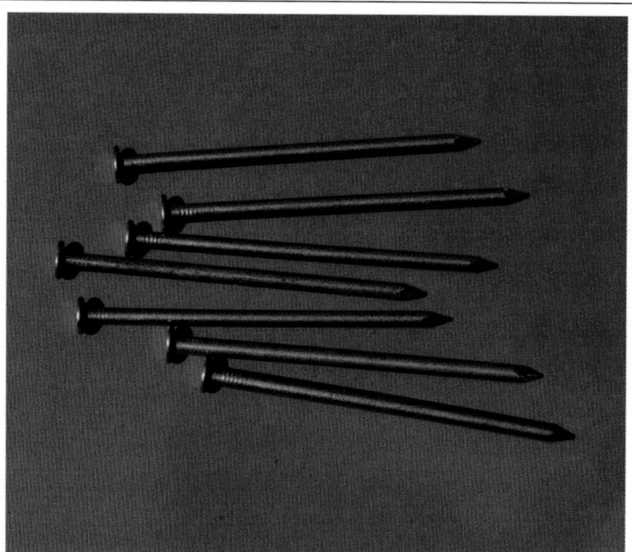

- Round-head nails are used when the nail heads can remain visible.
- Common nails are used with framing lumber, and box nails with narrower shanks (shafts) are used with thin pieces of wood.
- Bigger isn't always better—a large nail driven into thin wood will split it.
- Use nails about two times longer than the thickness of the wood being driven through.

out where you don't want it. Use too thick of a nail too close to the end of a board, and you can split it. Not certain what you need? Buy several sizes by the box. They store easily and always come in handy for other jobs.

For exterior work, always use galvanized nails to withstand moisture. Paint alone can't stop the head of an exterior nail from rusting.

ZOOM

Nails are referred to by "penny" size—2d or 2 penny, 4d, and so on. This is an old term referring to the amount of nails in a particular size that could once be purchased for a penny. A 4d nail is 1 1/2 inches long, an 8d is 2 1/2 inches long—the longer the nail, the higher the penny size.

Finish Nails

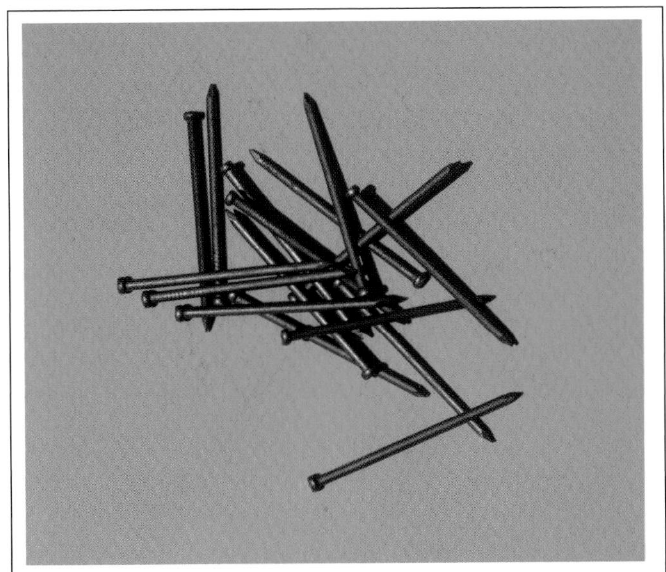

- A finish nail can often be used in place of a round-head nail, but you don't want to use a round-head nail for finish work.
- Finish nails can sometimes replace common nails if necessary—just use more of them.
- Brads are very small finish nails, usually sold in plastic tube-like containers.
- Wire nails are very thin brads but have larger, flat heads.

How to Nail

- Even experienced carpenters bend nails and hit thumbs. A few guidelines will keep your nails straight:
- To avoid splitting wood, blunt the sharp end of your nails. Place the head of the nail on a hard surface and tap the sharp end with a hammer.
- If your nails are a little long for the job, nail them in at an angle.
- Hammer with a few long, steady blows, keeping an eye on the head of the nail. Hold onto the nail only as long as you need to get it started.

CAULK

Wood, metal, brick, concrete, and tile—there's a caulk for every surface

Caulk is a flexible sealant. It keeps rain, wind, and insects out when it's used outside, and it provides a more finished appearance to woodwork and seals plumbing fixtures inside. There are dozens of types of caulk for an equal number of applications; latex and silicone caulks are the most common.

Each type of caulk comes in a long tube or cartridge, which fits inside a caulk gun. The guns have a movable pressure rod that presses against the bottom of the tube and forces the caulk out the nozzle at a controlled rate. Some are tedious to apply, but with practice, you'll get smooth results. Go slowly; it's easy to run too much caulk out and make a mess.

Caulk guns range from mediocre to excellent, depend-

Caulk

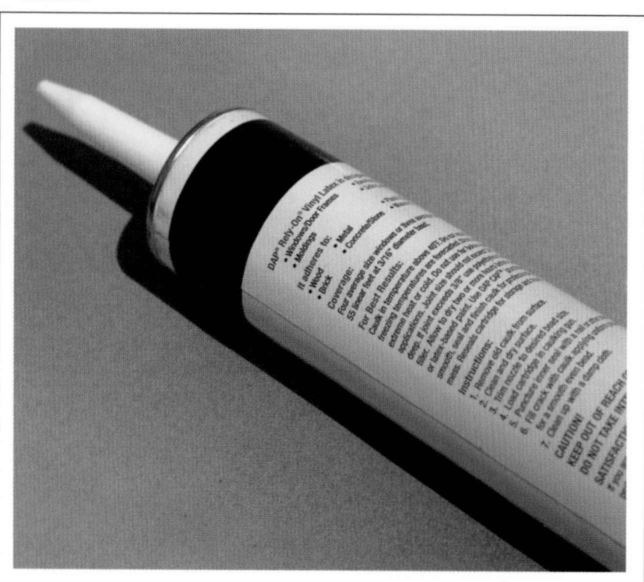

- Always read the label and technical information on a caulk tube before using it to assure you have the best material for your job.
- Most caulk will last for years after applied, but exterior caulk is more likely to need renewing from time to time.
- Acrylic latex caulk, a standard painter's caulk, comes in multiple colors, is used on wood, can be painted, and cleans up with water.
- All-silicone caulk is used on nonporous surfaces—tile, glass, metal—is very durable but not paintable, and cleans up with solvent.

Caulk Guns

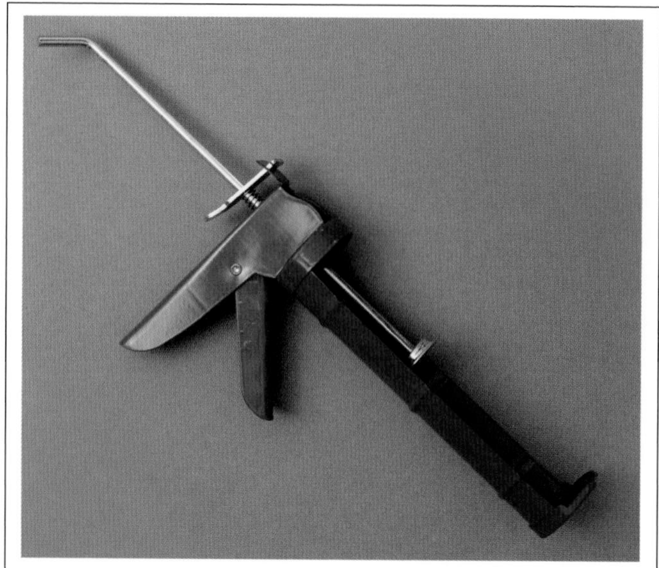

- Paint stores, which serve painting contractors, will normally have a good supply of professional-quality caulk guns.
- Better quality guns have a swing-out narrow metal cutter for puncturing the end of a caulk tube's nozzle and a thumb-operated lever release that stops the flow of caulk.
- When finished, remove the caulk cartridge and wipe down the gun.
- Some tub and tile caulk comes in small, squeezable tubes for quick repair jobs.

ing on their make and style. Look for an open-style caulk gun in at least the mid-price range at your home improvement, paint, or hardware store. The least expensive guns are the hardest to use and not worth the small savings. Before you caulk, the surface must be clean, mildew free, and dry. Remove any old, loose caulk. Cut the end of the cartridge nozzle at a 45-degree angle and just smaller than the opening you're filling. Apply caulk in warm, dry conditions.

YELLOW LIGHT

Old caulk past its expiration date should be tossed. Test any old caulk on a piece of scrap wood for flow, color, and drying time before using it in your project. Check with your trash hauler if the caulk can be disposed of in household garbage.

Applying Caulk

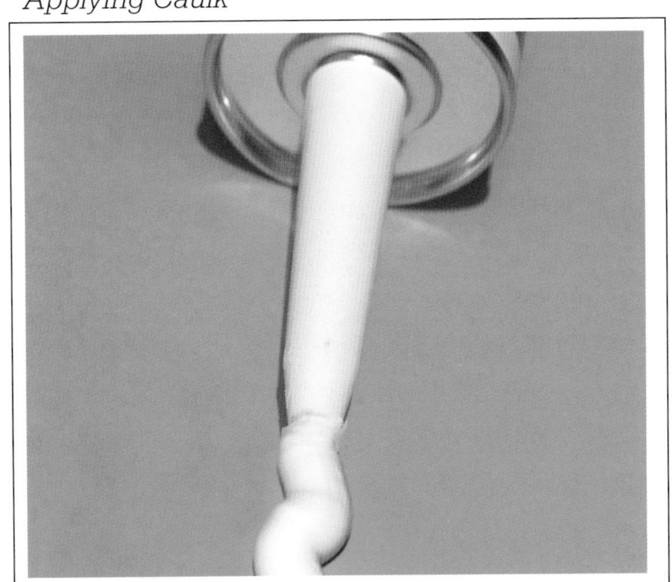

- Cut away any old caulk with a utility knife or putty knife before applying new caulk.
- Apply in warm, dry conditions, starting slowly while maintaining steady pressure as you move the gun, releasing the trigger near the end of the run.
- Run a finger or caulk spreader over latex caulk followed by a sponge to wipe up any excess (silicone caulk requires lacquer thinner).
- When finished caulking, wipe the end of the cartridge tip clean and press a long nail into it to prevent the remaining caulk from drying out.

Painting Caulk

- Latex caulk can be painted after it dries and cures. Siliconized latex caulk can be painted; pure silicone caulk cannot.
- Use paintable caulk for wood trim and next to most painted surfaces. Follow the curing times on the tube before painting.
- Clear caulk goes on white, dries clear, and is used when a sealant is needed that does not require paint.
- Applying paint to uncured caulk can cause the paint film to pull away and wrinkle as the caulk continues to dry.

TAPES & GLUES

Look beyond just duct tape and white glue to secure your projects

In home repair projects, tape is a unique material. It can secure plastic sheathing over a door opening to keep dust out, mask woodwork when painting, hold glued sections together until the glue dries, temporarily repair electrical cords and broken tool grips, and be used for marking just about anything when written on with a felt marker. A well-equipped toolbox will have rolls of masking, electrical, and duct tape in it for those odds-and-ends jobs that occur when you least expect them.

Paper tapes, primarily masking tape, have a limited life before the glue dries out and removing the tape becomes difficult. Blue masking tape is designed as a painter's tool to mask off painted areas; it lasts longer than plain masking tape because paint jobs can span several days, and it's less

Wood Glue

- White glue works well on porous materials not exposed to moisture, nonstructural repairs, and arts and crafts; it's nontoxic and dries clear.
- Use glue in addition to fasteners when repairing split doors or other woodwork for a tight, long-lasting fix.
- Yellow carpenter's glue or wood glue offers more water resistance than white glue; apply a modest amount to both surfaces and secure them together with a clamp or tape.
- Too much glue can actually weaken a glued joint.

Super Glue and Krazy Glue

- Cyanoacrylate is an acrylic resin in super-fast drying glues that cures almost immediately, secures all kinds of materials together, but isn't meant for construction purposes.
- Use this glue for odds-and-ends repairs involving glass, rubber, and ceramics.
- Brand name instant-bonding glues come in small tubes for a reason—you need only a little bit; any excess can be cleaned off with acetone (found in fingernail polish remover).
- These glues are not meant for outdoor use.

likely to pull off dried paint. Glue isn't a mechanical fastener in the sense of a nail or a screw. It's a chemical adhesive that holds things together. Some glues are designed to bind porous materials such as wood and paper, while others work best on nonporous ceramics. Water-resistant glue is a must on any surfaces near moisture. For many woodworking projects, standard yellow carpenter's glue does the trick.

As a rule, less glue forms a better bond than more glue.

RED LIGHT

Instant-bonding glue can accidentally stick your fingers together. Remove with acetone, found in fingernail polish remover. Apply a small amount on a Q-Tip and wash off immediately after glue is removed. An alternative is soaking the affected area in warm, soapy water until the skin can be rolled apart.

Contact Cement

- Contact cement is used to bond plastic laminate and many porous and nonporous materials but not wood.
- Contact cement is available in (flammable) solvent and water-based versions; use the flammable version in a well-ventilated area.
- Coat both surfaces with contact cement, which then sets up before the surfaces are pressed together.

Tape

- Repair jobs often call for a specific tape for a reason: Each is designed for certain purposes and is not always interchangeable.
- Vinyl or plastic electrical tape acts as insulation against the conduction of electricity, but it isn't meant to secure or strengthen anything.
- Masking tape is regularly used to tape off areas to keep paint off of them.
- Duct tape is often used as a sealant for ductwork, even though research suggests it's not effective for this use.

SANDPAPER & STEEL WOOL

Get familiar with abrasive materials that smooth, clean, grind, and polish wood and metal

Woodwork, floors, and furniture are normally sanded smooth before being finished, or the finish itself is sanded to remove any imperfections and to allow the next finish coat to stick better. When picking out your materials, be aware that sandpaper comes in many forms, sizes, and grades, that are distinguished by their grit (the small cutting particles glued to the paper). The higher the number—220, 280, 360—the finer the paper and the less abrasive its cutting.

Hand-sanding is appropriate for scuffing a finish, minor, quick smoothing, or for areas where dust is a consideration. Otherwise, power sanding with a sander is a far better way to go. You cannot duplicate the speed or movements of a

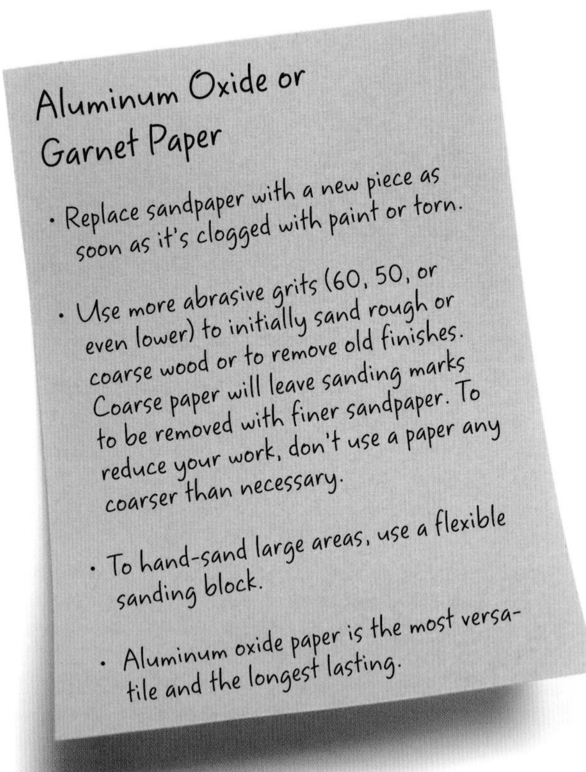

Sanding Sheets

- Sandpaper is sold in a variety of sheet sizes and packaging, from mixed packets of different grits to 50- and 100-sheet sleeves.
- Full sheets of sandpaper commonly measure 9" x 11" and can be folded and cut or torn into smaller sizes to fit sanders and sanding blocks.
- Some imported sheets are undersized and do not fold or cut evenly into smaller sections—avoid these.
- You can also buy paper presized for sanders, but it's more expensive.

sander with hand-sanding any more than you can outrun a car.

The material being sanded determines the grit of paper used. Tough jobs such as sanding off varnish call for coarser, heavier grits. If you use a grit that's too coarse, you can usually remove the sanding marks with finer grades of sandpaper.

For metal work, steel wool is manufactured in grades like sandpaper, but its coarseness is measured by the thickness of its steel strands.

MAKE IT EASY

Mixed packets of sandpaper or sandpaper by the sheet are fine for small sanding projects. For large jobs or ongoing jobs, such as sanding rooms full of woodwork and doors or floor sanding, consider buying by the sleeve (50 or 100 sheets) of the grit(s) you know you'll need.

Discs and Belts

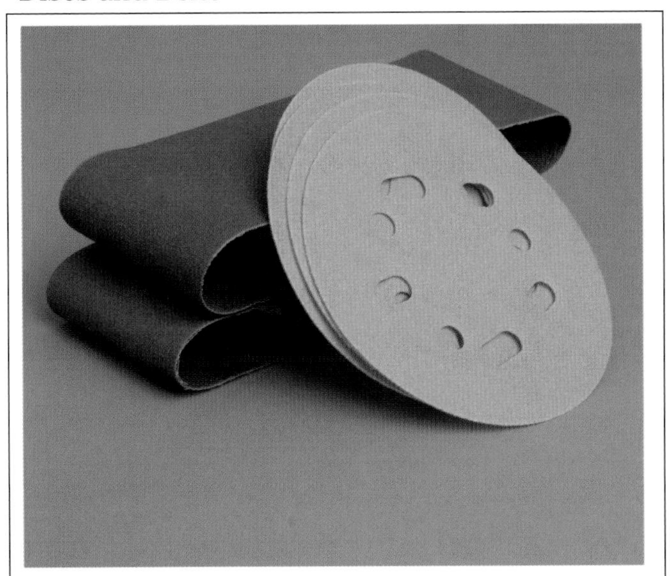

- Sanding discs range in sizes from 3/4-inch to 7 inches and larger and attach to a sander via hook and loop (Velcro), the most convenient to change, while others secure with a nut attached to the sanding pad.
- Be sure the discs you purchase will fit your sander—the wrong type won't work properly.
- Disc sanders cut fast—try a very light grit disc first to get used to the machine.
- Sanding belts range from 3/8 inch to 4 inches wide for commonly available belt sanders.

Steel and Brass Wool

- Steel wool comes in grades, its coarseness measured by the thickness of its steel strands.
- Extra coarse #4 removes some rust and corrosion, while extra fine #0000 can polish furniture with wax or bring a shine to brass.
- Less common brass wool is more resistant to corrosion when wet.
- Wear gloves to keep bits of steel strands out of your fingers. Discard wet steel wool when finished—it will corrode if you save it for another use.

PAINTS & FINISHES

Know your coatings: Latex, oil, clear, paint, and primer each have a different purpose

To protect wood, drywall, and plaster surfaces and make them cleanable, a coating of some kind must be applied. Unprotected wood exposed to the weather will deteriorate, possibly rot. Paint and other finishes seal surfaces and add color.

A paint store can be a very confusing place. Do you get satin latex or latex enamel? Wiping oil or spar varnish? Fast-drying primer or PVA? Any number of finishes can work on different projects, but you must decide on the degree of gloss or shine, the type of finish, and the ease or difficulty in applying that finish.

Most household paints are either latex (water-based) or alkyd (oil-based). Latex has been the overwhelming choice for

Primers

- Primer prepares a surface for paint; it is not a finish coat, nor is diluted paint a substitute for primer.
- Use latex primer with latex paint and oil primer with oil paint for the best results. On rough exterior surfaces, oil primer is recommended with latex paint.
- Use fast-drying primer for spot priming, not large areas.
- If exterior primer sits too long without being painted, the paint may not bond properly—follow the manufacturer's directions for curing times.

Paints

- Oil paint has a harder film, takes longer to dry, and fades faster than latex paint, and while it adheres well to different surfaces, it becomes brittle on exteriors and yellows with age.
- Latex dries quickly, remains flexible, allows moisture to pass though, and does not hold as well in high-traffic areas as oil-based paint.
- Oil works well on interior woodwork but less so on exterior siding or trim unless all previous coats of paint are oil-based.
- Oil-based paint is less sensitive to cold temperatures.

homeowners.. It's easier to apply and clean up, dries quickly, and forms a flexible film on exteriors where paint has to give a little and "breathe" as temperatures change and moisture passes through it. Paint gloss reflects its solids content: The higher the gloss, the more solids and the tougher the finish.

Primer gets applied first on unpainted surfaces. It seals and provides a faintly coarse surface for paint adhesion.

ZOOM

Oil paint sticks to most surfaces, is harder to apply than latex, has a more offensive odor, and a much longer drying time. Oil works well on interior woodwork but not on exteriors unless all previous paint coats are oil-based. Latex paint can be used over oil after thorough preparation and priming.

Different Glosses

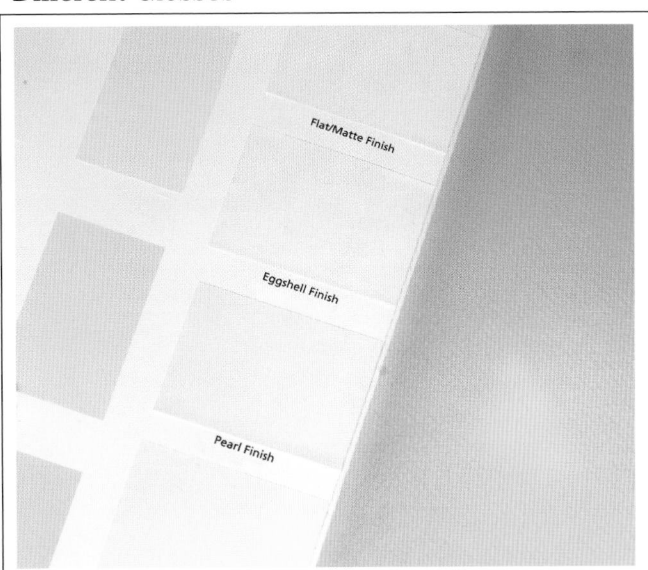

- Paint gloss indicates the solids content: the more solids, the tougher—and more washable—the finish. A higher gloss is shiny because it reflects more light.
- Flat paint absorbs light and hides wall and ceiling defects.
- Satin or eggshell's soft sheen is washable, making it appropriate for bathrooms, kitchens, and children's rooms.
- Semigloss paint was traditionally used on woodwork and shows imperfections easily, while high-gloss paint is used in marine applications and on furniture.

Exterior Paints

- Exterior paint with tougher resins and additional pigment withstands any weather conditions and covers all types of surfaces, including wood, stucco, shingles, and existing paint.
- Interior paint resists wear but can be used only on interiors.
- The highest quality exterior paint will be the least expensive in the long run, but even this paint will not hold up on a badly prepared surface.
- Dark exterior colors will fade more noticeably, and faster, than light colors.

STAINS

Stain colors surfaces and offers some protection but not as much as paint

Paint is loaded with solids, which obscure surfaces when paint is applied. Stain has a lower solids content and both penetrates and colors wood while allowing much of the grain to show through. Stain alone offers much less protection than paint, especially in an exterior application. It's common to see deteriorated siding and fences after one forlorn and distant past application of stain that goes for years without a recoat. You should expect to recoat exterior stains twice as often or more than paint. On the plus side, exterior semitransparent and transparent stains don't require a primer coat and are easy and quick to apply. They do not build up a surface film and never blister or peel as paint can.

Interior Stain

- A small amount of stain goes a long way—you can somewhat control stain color by how much you apply and how fast you wipe the excess off.
- Stain color looks different on different woods.
- Test your stain color in both daylight and nighttime lighting conditions to be sure the color is what you're looking for.
- Apply stain to wood scraps before using in your project.

Exterior Stain

- Semitransparent or transparent exterior stains require more frequent recoating (every 3–5 years depending on weather exposure) than paint; solid-body exterior stain is like a light-bodied paint but without the holding power.
- Solid-body stains spread more easily than paint but don't form as strong a film.
- Stains contain fungicides to fight exterior mildew.
- Oil-based stains penetrate deeper than latex stains and require clean-up with paint thinner.

Interior stains are always followed by some type of clear coating to protect and seal them. Otherwise, the stains will wear off after foot traffic or other use. Premixed interior stains are easy to apply with a brush, rag, or combination of the two and come in a variety of colors. Oil stains are considered to penetrate deeper than latex and have sharper color.

A clear exterior water-repellent preservative offers limited protection for about a year. Some painters apply this before applying an exterior stain, but not all clear preservative manufacturers recommend this.

MAKE IT EASY

Always test softwood such as fir and pine for stain absorption. It's often necessary to seal interior softwoods first so the stain goes on evenly. After testing your stain color, consider going one shade lighter. Stain selection can be more problematic than paint color, and going lighter is often the right choice.

Applying Stain

- Interior stains are brushed onto woodwork or applied with a rag to furniture, and both are wiped down with clean rags to remove any excess.
- Exterior stain is brushed, rolled, or sprayed and is absorbed quickly into raw wood.
- Stain the entire length of board or dimensions of a furniture section; stopping partway and picking up later can leave an irregular appearance.
- Maintain a wet edge as you stain, brushing from a dry area into the previously stained wet material.

Clear Water Repellents

- Clear water repellents contain polymers or wax and offer some protection, but an extra coat of stain probably provides more.
- Always follow the manufacturer's application instructions—these products can become tacky and not dry properly if misapplied.
- Don't mix a water repellent with an exterior stain—apply each separately.
- Water repellents also protect and help seal masonry but are not normally recommended on historic structures and can discolor brick (always test first).

CLEAR FINISHES

From traditional varnish to modern lacquer, clear coats protect and shine through

Clear finishes are broken into two broad categories: those that form tough surface films and those that form weaker films. These categories include oils derived from plants, varnishes, polyurethane, lacquer, shellac, and modified oils. Pure oils penetrate and form a soft film, which offers mild resistance to wear and tear but is easily recoated with a clean rag. Varnishes and polyurethane also penetrate, form a harder film, and are not as easily renewed or reapplied as oils.

Oil-based and latex clear finishes protect stained and unstained wood and allow the wood grain to show. Flat, satin, and higher gloss clear finishes are available, although the softer-appearing satin is commonly used in homes.

Varnish and Polyurethane

- Polyurethane and varnish each form a clear, durable film, although water-based polyurethane will appear milky when first applied.
- Apply in 65-degree temperature or higher to avoid drying problems, especially with oil-based products.
- Clear exterior finishes, sometimes seen on entry doors, need recoating more often than a painted finish.
- Apply at least three coats of these finishes over raw wood for complete coverage and penetration, carefully following the manufacturer's drying time requirements.

Oils

- Pure and modified oil finishes require more frequent application than varnish and polyurethane, depending on the application, because they form weaker films.
- Pure tung oil is expensive, requires a lot of rubbing, is easily repaired and maintained, and is slow drying.
- Modified oils, also known as an oil/vanish blend, are sold under a variety of brand names and are usually applied with a rag.
- Wiping varnish, another finish, forms more of a film than modified oil.

Pure oils include tung and linseed. Both are used as ingredients in other finishes, including varnish and modified oils. Tung oil takes some effort to apply and is very slow drying. Popular modified oils have added resin and paint thinner, are easy to apply, offer moderate protection, and require multiple coats.

Shellac is a favorite with some antique restorers. It offers little resistance to moisture and comes in various colors. However, it dries very quickly, so it can be difficult to apply. Shellac can be an undercoat for varnish and sticks to all kinds of surfaces.

ZOOM

Lacquer, another clear finish, dries extremely fast and is often used as a furniture finish on interior woodwork and prefinished doors. Brushing lacquer is also available, but varnishes and polyurethane offer harder finishes. Spray cans of lacquer work well for small art projects but are inappropriate for larger jobs.

Shellac

- Shellac is very much a specialty finish used by restorers and hobbyists and makes an excellent stain blocker in shellac-based primer/sealers.
- Shellac has no petroleum-based ingredients and no paint thinner odor.
- Water will leave a white mark on a shellac finish that has added wax.
- Note the expiration date on any can of shellac.

Wax Finish

- Wax alone is not effective as a final finish, but it maintains furniture finishes as a top coat, brightens metal, and offers some stain protection to worn kitchen sinks.
- Liquid wax is easier to apply but offers less protection than paste wax.
- Apply wax sparingly—otherwise, too much is wiped off and wasted.
- After applying, wax solvent begins evaporating, and the wax turns hazy. If the wax sits too long and is hard to rub off, apply fresh wax to resoften.

RECYCLED PAINT

Mix your own or buy manufactured recycled paint for a "green" finish

It has been estimated that Americans buy 650 million gallons of paint each year. As many as 35 million gallons become surplus, another way of saying they're ready to get tossed into the trash. Local municipalities are increasingly offering collection programs for leftover paints and other household chemicals, but all at a price. Paint represents just over a third of collected household hazardous wastes and costs almost $8 a gallon to manage, transport, and reprocess as a recycled product. The market for recycled paint is minuscule but growing slowly. Until consumers understand recycled paint is manufactured to the same quality standards as new paint, the market will remain small.

Recycled paint is a quality-tested product and a bargain

What Is Recycled Paint?

- Recycling paint offers an opportunity to greatly reduce landfill waste and consumer paint costs.
- Usable, but unused, paint is mixed, new materials are added and tested for quality assurance in a limited range of colors, and the paint is sold as a guaranteed product.
- Municipal recycling departments have information on locally available recycled paint and paint manufacturers.
- Recycled paint is less expensive than new paint and is gradually expanding in the marketplace as a viable product.

Mixing Your Own Paint

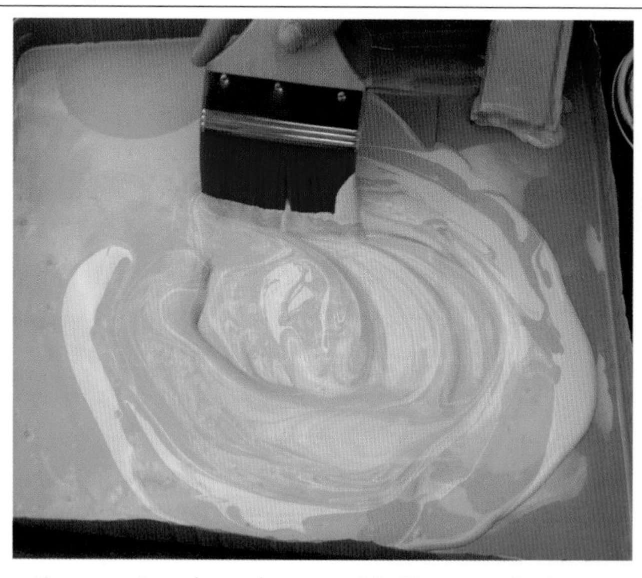

- You can mix and recycle your own leftover paint, too, but avoid very old paint that might have mold in it or is too stiff to liquefy.
- Be sure you don't need any of your leftover paint for touch-up.
- Mix like types of paint—interior latex with latex—together in a clean 5-gallon bucket or suitable container.
- This paint is perfect for garages and basements but usable in other rooms, too.

compared with new production paint. It isn't a new idea, either. Recycled paint has been around since the early 1990s. However, the color selection is limited, but workable, and availability varies from city to city.

You can create your own recycled paint by mixing any leftover, unneeded paint of like type (latex and latex, for instance) in clean containers. The resulting color might be unexpected but perfectly usable for a garage, basement, or storage room. As long as the mixed paints are not terribly old or don't have mold present, your final product should be perfectly usable.

GREEN LIGHT

Recycled paint keeps usable paint out of landfills. Paint companies are gradually producing lines of recycled paint, a sure sign their use will grow. Colors may vary from time to time, so buy enough for touch-up later. Recycled paint undergoes the same quality assurance during manufacturing as new production paint.

Paint Colors and Quality

- Recycled paint cannot offer the same color range as new paint since it's dependent on whatever paint is available for the mix; most common colors will be beige, shades of white, grays, and brown (check with your recycled paint dealer for color availability).
- Product quality is assured, but if you have any questions or concerns, discuss them with your paint supplier.
- Various government agencies have tested and approved of recycled paint.

VITAL PRODUCTS

FAUX FINISHES

Create a stone, marble, or wood grain finish using paint and a few simple tools

A faux finish is a disguise that turns common drywall or woodwork into an exciting and enticing image. Trompe l'œil or "trick of the eye" is a type of faux finish that creates the optical illusion of depth in a painted image. Faux finishes can range from techniques in applying ordinary house paint to the use of special glazes, plaster, and associated finishes to obtain a traditional faux look.

Like many decorating techniques, faux finishes, which date back thousands of years, have periods of popularity and rejection. After a resurgence in the early 1990s, faux finish materials and tools are readily available for all skill and idea levels.

The beauty of a faux finish, aside from its intrinsic appear-

Compwood Graining

- Wood graining renders a softwood or less expensive wood to resemble a hardwood. For instance, pine is grained to resemble oak.
- It is hard to distinguish a successful graining job from real wood.
- Raw wood is sealed with a base coat, including shellac or paint, and then a glaze finish is applied and worked with a series of tools to produce a desired wood grain appearance.
- When dry, the grained finish is coated with a clear finish for protection.

Marble Finish

- Marbling can be done to any woodwork, floors, and walls, but too much can overwhelm a room.
- Color mixing and patterns are critical in good marbling—expect to practice first before doing your final work.
- To get a realistic effect, the different shapes and addition of veins in the finish are multiple tasks that can't all be done at once.
- Some protective clear coats will yellow with age, affecting the appearance of the marbling.

ance itself, is the ease of changing or painting over a project that just didn't work out. With practice, an amateur can get very presentable results. One approach for this practice is to do the inside of a closet or laundry room until you're satisfied with your technique and tool selection.

Virtually anything can be used to tool a faux finish: rags, sponges, paint rollers, common and special brushes, combing tools, and feathers for marbleizing. A small investment in time and money can yield surprising results.

MAKE IT EASY

Internet sites are fine for introductory faux finish information, but a book offers far more information, techniques, and photos of sample finishes all in one place. Check out the Resources section (page 228) or your public library for more information on faux finish books and websites.

Common Graining Tools

- A professional faux finish specialist has an array of brushes and hand tools for specific effects, but you can get by with fewer.
- Sea sponges, cotton rags, small pieces of burlap, chamois cloth, and cheesecloth are all used to produce marble and wood grain effects.
- Aside from brushes to apply and spread the basecoat, use separate brushes with both coarse and fine bristles for drawing out wood grain, creating marble veins, and dragging a finish.
- Steel combs are used for creating wood grain.

Faux Finish with Basic Tools

- There are plenty of online and other suppliers of faux finish tools and materials, but you can also create some interesting finishes with what you have.
- Kitchen sponges, wool socks, plastic pot scrubbers, and damp wash cloths can duplicate various faux effects.
- Branches from shrubs, especially evergreens, can be dragged through a wet glaze finish for surprising results.
- Children's paint sets have the ideal-size paintbrushes for some veining techniques.

FILLERS

Holes, cracks, and damaged surfaces all need fillers to complete their repairs

Fillers, like so many other materials, come in all forms, but they all do the same thing: They fix imperfections in surfaces before painting. Some are very dense and strong, while others are softer and more pliable. Some dry very fast, which means you have only a short time to work with them, while others are more forgiving. The most common filler for small holes is Spackle, which despite being a trademarked name has become so universally used it's become a generic term.

Some fillers have migrated from their original purpose to home repairs. Automotive body filler, generically referred to as "Bondo," another trademarked name, can fill large holes if applied in several applications. It does not shrink when prop-

Spackle

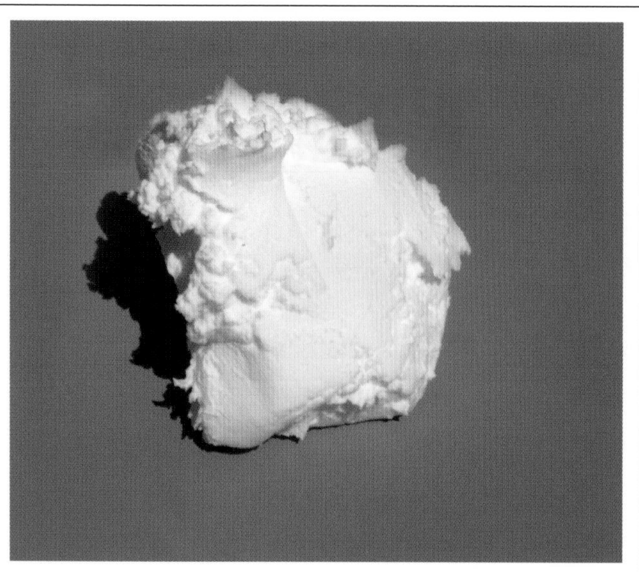

- Spackle is a premixed, ready-to-use pliable plaster-like paste filler sold in interior and exterior versions.
- The name Spackle is a trademark that has become so universally used it's become a generic term for a prepainting filler.
- Spackle fills only small holes and should not be used for large repairs—it won't dry properly, and the repair will fail.
- Keep Spackle sealed in its container and discard if it dries out.

Automotive Body Filler

- Automotive body filler is a two-part mix that dries hard in minutes, holds paint well, and should be used sparingly on finish work.
- It comes with a separate hardening agent, but buy an extra tube of the hardener because it often runs out before all the filler is used up.
- Automotive fillers come with or without reinforcing fiberglass strands for extra strength and can be shaped and sanded to replace missing corners or decorative plaster.
- Do not nail or screw into automotive body filler.

erly applied and can be shaped and sanded to replace missing corners or decorative plaster.

Traditional plaster work consists of three coats of different types of plaster, so patching plaster repairs all three layers with easy-to-mix filler that must be worked quickly before it dries. For deep holes, apply it in multiple, thin coats.

Wood filler or wood putty fills holes and pores in unpainted wood. It's formulated to match the appearance of wood so it won't stand out too much when stained or clear-coated. Wood fillers dry quickly and sand easily.

MAKE IT EASY

Buy a larger box of patching plaster than you think you'll need for one job because you'll always need more later. Until you're used to mixing the correct amount for the job and working it before it dries out, you'll experience some wasted material.

Plaster Patch

- Plaster patching fillers are packaged in powder form, get mixed with cold water, set up to a stiff consistency in a few minutes, and dry hard shortly thereafter.
- Use just enough to do a smooth job; once dry, it's very difficult to sand.
- Use Spackle or other soft fillers as a final finish coat over small repairs.
- Patching plaster can also be used to repair and fill wood holes and cracks, as well as some floor repairs prior to installing vinyl or carpet.

Wood Filler

- Most wood filler, a combination of a binder and wood flour or similar fine material, is sold premixed in cans in both solvent and water-based formulas.
- A little goes a long way—use it sparingly.
- Wood filler comes in a variety of colors to mimic the natural appearance of some woods, although no match will ever be perfect.
- Open cans dry out quickly—keep the lid on except when removing the filler.

SCRATCHED WOODWORK

Love your pets but hate their scratches on your furniture? Here's the fix you need

As much as they would like to, cats and dogs can't open doors, but they can scratch at them and at other woodwork. Several of the fillers previously mentioned can repair these scratches. Your choice of filler depends on the depth and number of scratches, the detailing of the woodwork, whether it's painted or stained and clear-coated, and the amount of dust you're willing to put up with.

Light scratches can be filled with the door in place—use a drop cloth or piece of plastic on the floor—or removed and placed on sawhorses elsewhere. Heavy scratching calls for more filler and sanding, which can be easily done with a power sander on the removed door. Usually the surrounding

Pet-Scratched Door

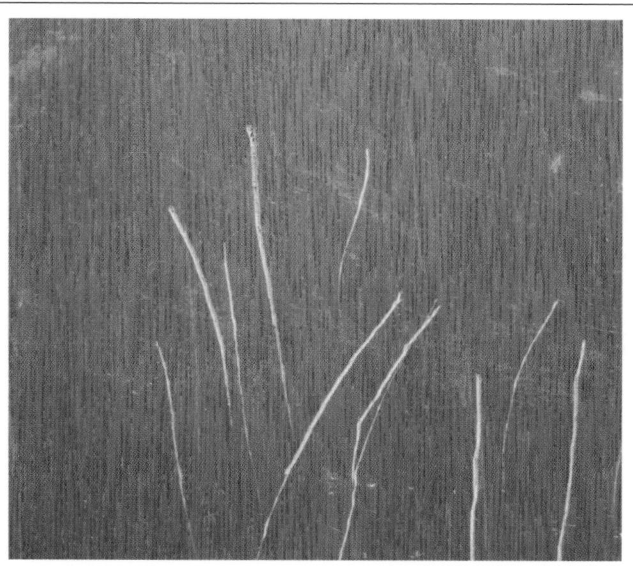

- Spackle covers cracks up to 1/8 inch or so and is easy to sand.
- Automotive body filler (see page 46) can fill almost any size opening but is difficult to hand-sand if applied excessively.
- Don't waste wood dough on painted wood—it's meant for unpainted wood repairs and is too expensive for repairs that will be painted over.
- Patching plaster is fine for wood repairs but not as easy to smooth out as Spackle.

Filling the Scratches

- Lightly sand the scratched area, wipe off the dust, press the filler in with a clean putty knife, and allow it to completely dry.
- Keep the amount of filler to a minimum to avoid extra sanding.
- Wipe the putty knife clean when finished to keep built-up filler from drying on the edge.
- Stained woodwork and doors can be touched up with a matching wax stick if the scratches aren't too deep.

woodwork is less scratched and requires less filling.

Hand-sanding filler throws out the least dust. Machine-sanding removes excess filler faster, but the clean-up can be a real mess, especially if the door and woodwork are sanded in place.

The condition of the repaired and surrounding areas determines how much work is required. Paint a complete section, even on recently painted wood. A complete section blends in better than trying to touch up just the damaged area.

RED LIGHT

If the door or woodwork is painted with lead-based paint, you must take precautions against spreading contaminated dust. The precautions are considerable if a power sander is used, especially if children, pregnant women, or pets live in the house. See the Resources section on page 228 for more information.

Sanding Fillers

- Wear a dust mask and start with 80-grit sandpaper to sand most fillers smooth.
- Sand in the direction of the wood grain if sanding by hand.
- If the door is removed, sand the patched area with a palm sander or a random-orbit sander until the area is smooth; follow up with 100-grit paper.
- Consider a final sanding with a finer-grit paper—120 or 150—and wipe off all dust before applying the finish.

Priming and Painting

- The repaired and surrounding areas determine how much repainting or staining/clear-coating is required.
- At a minimum, coat the complete section of a door containing the damaged area and recoat the entire section when woodwork is repaired.
- Limit recoating to a logical area—all the trim around a door, for instance—if one section is damaged and recoating it alone will stand out too much.
- Sometimes a stained door can be touched up acceptably; otherwise, sand out that section and refinish.

SCRATCHED FURNITURE

Repair a few scratches without refinishing the entire piece of furniture

Wood furniture is subject to nicks, scratches, and spills. Some can be repaired without stripping and refinishing. Because furniture has so many distinct sections—legs, individual drawers, tops—you can repair any one of them individually to match the rest of the piece. This is what good antique restorers do all the time to repair minor damage.

Start with the least invasive approach—in some cases, cleaning and polishing the affected area with a polish/wax combination and very fine steel wool. In others, a putty stick, which is like a specialized wax crayon for filling scratches on stained furniture, produces an acceptable repair.

The furniture finish will influence the repair. Refinishing a

Scratched Furniture

- Fill small cracks, nicks, and scratches with color-matching finish putty or a wax stick.
- Paint markers and paint pens, available at some fine furniture stores, contain dyes for blending in scuffs and scratches.
- White rings can often be removed with denatured rubbing alcohol on a clean rag—go easy, you don't want to dissolve the old finish—followed by furniture wax.
- Chips in a clear finish can be carefully touched up with clear fingernail polish.

Damaged Finishes

- Sunlight can cause a clear finish to shrink and crack, and it can leave furniture with a bleached-out look.
- Many pieces of furniture have veneered sections—thin sheets of expensive hardwood glued to less expensive woods.
- Veneer can be carefully sanded with light sandpaper to remove an old finish, but furniture refinisher is safer.
- Furniture refinisher is available anywhere paint is sold and at many antique shops.

table top with a lacquer finish over stain requires scrubbing all the lacquer out of the wood before restaining. Otherwise, the new stain can appear splotchy. A wax finish over oil is simpler to work with.

Sunlight also damages furniture finishes, especially on top. If the rest of the piece looks acceptable or can be buffed up a bit, the top alone can be refinished.

Whenever any of the old finish is completely removed, replace it with new stain and a clear top coat to match the rest of the piece and protect the wood.

MAKE IT EASY

Furniture refinisher, a combination of solvent and oils, is suitable for spot refinishing on a piece of furniture. Heavy-bodied paint remover is less expensive and more versatile overall than furniture refinisher but can be overkill for simply stripping a furniture top. The simplest repair is to cover a damaged top with something decorative.

Staining Furniture

- An alternative to stripping is to carefully paint on new stain; the finish will be darker, but the job is easier.
- Lightly sand the damaged and surrounding areas with 120 and 150 sandpapers, and wipe away the dust.
- Brush on just enough matching oil stain to evenly cover the sanded area.
- Run the brush tips across the recoated area very lightly to even it out and allow it to cure for three days; then coat with varnish or polyurethane.

Clear Coat

- When recoating woodwork with a clear coat, try to match the type of material used on the surrounding woodwork and its sheen; on a furniture top, apply whatever coat will stand up best depending on how the furniture is used.
- Multiple coats of a modified oil finish are the easiest to apply and touch up.
- Apply three coats of varnish or polyurethane over bare wood, no fewer.
- Fast-drying lacquer is available in convenient spray cans.

WOOD FLOOR SCRATCHES

Touch up your wood floors and skip a complete refinishing with these simple tips

Wood floors look beautiful when they're new and freshly finished, but beauty can be fleeting as family life takes over. Wood floors get scratched by the usual culprits: grit on shoe bottoms, pet paws, dragging chairs, and kids' toys. In the day-to-day family life, scratches tend to be unavoidable.

Small scratches can be repaired the same as furniture scratches. The depth of the scratch, in part, determines the repair. Deeper single scratches might need sanding out.

The key is not to overdo the amount of sanding but also to sand out enough, sometimes the complete length of a floor board, in order for the touched-up area to match the surrounding finish and not stick out. The problem with attempt-

Sanding

- Tape off the edge of the entire length of the affected floor board(s) with blue masking tape.
- Hand-sand (or machine-sand if dust isn't a problem) the scratch out with 80- and 100-grit sandpapers, being careful to stay inside the tape boundary, finishing with 120-grit paper.
- Sand the finish off the rest of the board(s), and vacuum and wipe off all the dust.
- Apply appropriate floor finish coats with foam brushes.

Sanding a Larger Area

- A floor can have multiple scratches in the top coat over a large area that looks unappealing but doesn't justify a complete refinishing. The area can be recoated for an acceptable appearance and added protection.
- Sand the scratched area with 120-grit and 150-grit sandpapers, and vacuum and wipe away all dust.
- An existing oil-based finish can be softened with lacquer thinner.
- Recoat with the same finish—don't mix oil-based with water-based—following the grain of the wood.

ing to clear-coat a repaired area in the center of a board, for instance, is the gloss never quite matches the surrounding finish.

If a large area of the floor has multiple but superficial scratches, it's appropriate to recoat a large area after lightly sanding and preparing the scratched area to receive an additional coat of finish.

Of course, the easiest repair is placing a rug over the damaged area until the entire floor gets refinished.

YELLOW LIGHT

Be sure the new finish used to touch up a scratched area is compatible with the existing finish. Water-based should go on water-based, oil-based on oil-based, etc. Applying wax can cause refinishing problems later if it seeps into the wood pores and isn't completely removed.

Clear Coat

- While wearing a respirator, apply an oil finish with a good-quality bristle brush or a pad applicator; the brush gives you a bit more control.
- With either an oil- or water-based finish, spread evenly along the entire length of two or three boards.
- Work quickly to maintain a wet edge as you apply from one section to the next.
- Follow drying instructions, noting when you can use the floor again.

Floor Refinisher

- Wood floor restoration products contain some solvent and wax or polymer-type finish and are used to clean and revive scratched and worn floors, but it's not the same as refinishing.
- This is a maintenance approach until a floor needs a complete refinishing.
- Used regularly, floor restore/reviver products can be a good alternative to refinishing.
- Various products are available at tool rental shops and online under "hardwood floor restorer."

SCRATCHED COUNTERS

Learn to buff out and repair scratches in even your toughest laminated countertops

Laminated countertops are made from layers of plastic-coated paper pressed together at high temperatures and extreme pressure. They're mostly waterproof and moderately resistant to scratches. Plastic laminate has its roots in electrical insulators created in the early twentieth century. This material is available in a multitude of colors, patterns, and glosses and is the predominate countertop material in new kitchens and bathrooms.

Laminates vary in their composition. Some are colored all the way through and others only on the surface. Scratches on the latter are more noticeable. Depending on the depth of the scratch or burn, you might be able to buff them out.

Gel Gloss

- Sharp edges will damage plastic laminate, as will heavy metal objects moved across or dropped on it.
- Hot pans and abrasive cleaners also damage plastic laminate.
- Gel Gloss, a combination of nonabrasive cleaners and wax, removes many stains, rubs out and obscures some scratches, and adds a gloss appearance to plastic laminate.
- Dark colors fade and show damage more readily than textured and lighter colors.

Felt Pen

- Use a felt tip marker as close to the laminate color as possible.
- Draw the tip over the scratch—if it's not a good match, erase it with a bit of lacquer thinner on a rag.
- Let a good match dry, and wipe away any excess ink around the scratch with lacquer thinner.
- Go over the entire counter with Gel Gloss or rub some automotive paste wax into the scratch.

Some commercial products clean plastic laminate and fill the scratches with a liquid polymer that can be renewed from time to time. These products do not eliminate the scratches but make them less noticeable.

With professional help, plastic laminate can also be covered with new laminate, or countertops can be completely replaced with new material, an expensive proposition. For large burns that can't be buffed out or glossed over, cover and conceal them with another counter material until the entire counter gets redone.

MAKE IT EASY

After it's colored with an appropriate marker, a scratch can also be filled with carefully applied clear nail polish, wiping away the excess with lacquer thinner or nail polish remover. Wax or Gel Gloss can also be used, but the nail polish or similar clear coat will last longer.

Butcher Block

- You might be able to buff some faint burns out of laminate with polishing compound or #0000 steel wool and some Gel Gloss, but more prominent burns cannot be cleaned out, although they can be covered up.
- Clean grease from area and sand it with 120-grit paper.
- Glue sections of marble, butcher block, or heat-proof tiles over the burned area.
- Use just enough waterproof glue or adhesive to hold the material in place.

Install New Laminate

- Installing laminate takes experience, specific tools, and know-how. Measurements can't be changed once the new laminate is glued into place. For many, it might be best to call an expert.
- Remove the sink and measure the existing counter. Then sand and wipe it free of dust.
- Cut the laminate to fit and glue it down with contact cement.
- After the adhesive dries, trim the laminate and cut a hole for the sink.

REPAIRING A CHIPPED TILE

Repair chipped tiles quickly and easily without the help of a professional

Tile is a tough countertop, floor, and shower stall material and can last for generations. Both ceramic and porcelain tiles are made from different types of baked clay, with porcelain tile being the stronger of the two. Glazed finishes, essentially liquid glass, seal the tile to water and stains. As water resistant as tile may be, its weak point is its grout, a cement-like material that seals the joints between individual tiles.

Tiles are installed in wet areas—bathrooms, kitchens, and mudrooms—and in dry areas such as fireplaces. A chipped tile in either environment can be repaired, but wet areas require more diligence. Chipped and cracked tiles can be replaced with identical tiles if spares are available or with a contrasting

Using Epoxy Filler

- Clean out the chipped area completely, digging out loose material with an old screwdriver.
- Prepare a small amount of paintable, waterproof epoxy.
- Use small pieces of cardboard covered with aluminum foil to keep the epoxy within the damaged area (some epoxies will self-level, and others require smoothing out with a putty knife).
- When it is dry, touch up with an oil paint (see page 36) to match the color of the tile.

Removing a Tile

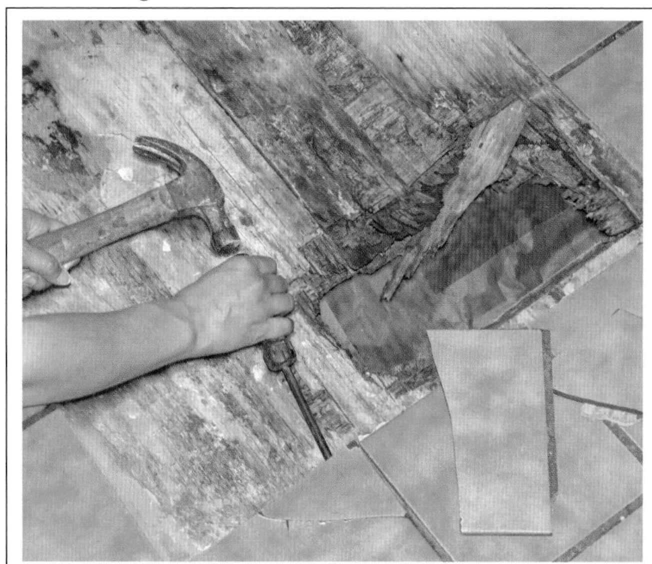

- To remove a tile, first, carefully remove the old grout (wear safety glasses).
- Cut through the grout by using a grout saw, available at hardware stores and home centers, or utility knife or by drilling a series of small continuous holes. Scrape out any remaining grout with a stiff putty knife or old screwdriver.
- To ease removal, drill a series of short holes through the tile using a tile or ceramic cutting bit.
- Chip out the tile with a cold chisel.

color tile. Repairs can also be done with waterproof epoxy and oil-based paint in a color matching the surrounding tile.

As strong as tile is, damaged sections must be carefully removed to avoid damaging adjoining tiles. It's easy to do and frustrating when you find yourself replacing an additional tile or two beyond your original repair.

Don't expect to find exact matching tiles to old patterns or colors. Consider an artistic alternative with something colorful or fun as long as you can live with the result.

MAKE IT EASY

A chipped fireplace tile can be repaired with automotive body filler if epoxy isn't handy. Shape and sand the filler to match the missing area and paint when dry. Coat the paint with a bit of polyurethane to match the sheen on the surrounding tile.

Installing a Tile

- With the broken tile removed, scrape away any old adhesive and wipe with a moist rag.
- Check the new tile for fit and match and spread tile adhesive on the back of the new tile to within 1/4 inch or so of the edges.
- Line the tile up with the surrounding tiles and press and hold in place.
- Secure the tile with masking tape and let dry according to the adhesive manufacturer's instructions.

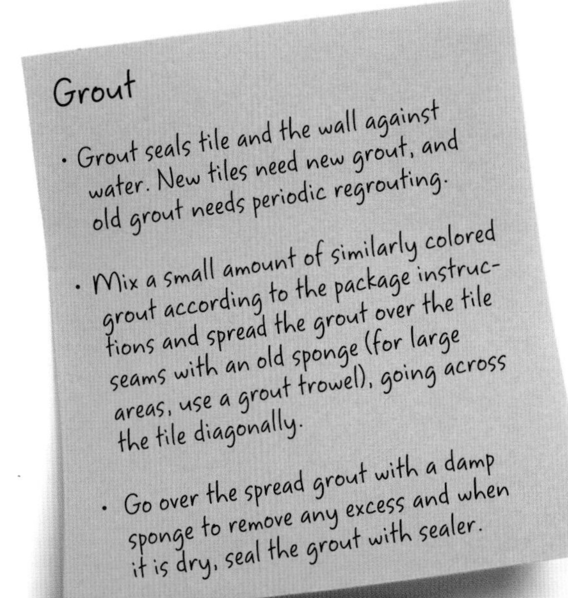

Grout

- Grout seals tile and the wall against water. New tiles need new grout, and old grout needs periodic regrouting.
- Mix a small amount of similarly colored grout according to the package instructions and spread the grout over the tile seams with an old sponge (for large areas, use a grout trowel), going across the tile diagonally.
- Go over the spread grout with a damp sponge to remove any excess and when it is dry, seal the grout with sealer.

REPAIRING A CHIPPED FINISH

You can't reproduce a factory finish, but you can still repair one

Appliances are coated with some of the toughest paint around. Special touch-up paints are available with brushes attached to the bottle tops, but they will never match the durability or finished appearance of a factory coating. That said, no one wants to look at a paint chip, so even a less-than-perfect repair is better than none at all.

Metal sinks and metal tubs have baked porcelain and ceramic finishes. These can chip if something heavy falls on them in just the right way and at just the right spot. Very old cast iron and steel tubs often show stains and worn finishes that can be renewed by any number of bathtub refinishing processes, although none that uses sprayed-on epoxy and

Applying Epoxy Paint

- Scuff the chipped area with 120-grit sandpaper, feathering the rough paint edges, and then wipe away all the dust.
- Apply two thin coats of appliance touch-up paint, available at appliance dealers, following the manufacturer's instructions.
- Depending on the manufacturer, this paint can sometimes be used on plumbing fixtures.
- Appliance touch-up paint is not long-lasting around water or high heat and will need recoating from time to time.

Porcelain Repair Kit 1

- Clean and lightly sand the chipped area, feathering the rough edges.
- Follow the instructions and mix just enough epoxy solution to fill the damaged area.
- Working quickly, as the material stiffens quickly, but carefully brush in one thin coat and allow the epoxy to self-level, following with a second coat if needed.
- This repair will never be as strong as the original finish but will do an adequate job of covering a chipped area.

other exotic paints will equal the original baked on porcelain. Some processes are better than others, and any consideration for using one of these services should be thoroughly researched first.

For in-home repairs of small chips, two-part epoxy solutions are more durable than premixed epoxy paint. They do not require any follow-up sealant and should be monitored for potential recoating. Some of these epoxy solutions are sold in small repair kits for touch-up and are available in popular appliance colors.

MAKE IT EASY

The key to touch-up repairs for fixtures and appliances is to completely clean the area being repaired according to the paint or epoxy instructions. It must be dry and undisturbed while the coating cures. If the chip is deep, a second coat might be needed to match the surrounding surface.

Porcelain Repair Kit 2

- Although other touch-up and repair kits are available in various colors to match popular colors, Porc-a-Fix is premixed in multiple colors by manufacturer, including older colors no longer available.
- Premix assures proper adhesion and longer-lasting repair.
- Color matching means the repair is less likely to stand out, but there will still be some difference in its appearance.
- Prepare the damaged area according to product instructions and apply, observing all drying times.

Chipped and Worn Faucets

- There are expensive vintage replacements for worn faucet and tub spouts, but you can do passable repairs with some solder, a metal alloy that, when melted, joins other metals together.
- Clean the corroded area with some fine sandpaper and wipe away the dust.
- Using a soldering iron and soldering wire (solder in wire form), gradually fill the area with a small amount of melted solder, allowing it to level out.
- When the area is full, polish the solder down with #000 steel wool.

FAUCETS

Stop dripping faucets and save both water and your sanity with these simple steps

Dripping water—slow, random dripping water—shows up in movies from time to time as a way of driving one of the characters crazy. And it also drives homeowners crazy. Stay sane by stopping these drips quickly and easily.

Without faucets, water would shoot through the open ends of pipes. A faucet allows you to control how much water comes out at any given time. Inside the faucet is a rubber or plastic washer or its equivalent. This washer compresses when the faucet handle is in the off position. Eventually, a washer needs replacement. It lets you know by allowing a small amount of water to drip through the faucet even though it's shut off.

Dismantling a Single-Handle Faucet

- Single mixing handle faucets are common in modern kitchens; each manufacturer uses a different cartridge or adjuster.
- Before replacing either a cartridge or ball assembly (don't bother with individual parts, just buy the whole assembly), shut the water off using the shut-off valves under the sink.
- Push the handle back and with an Allen wrench (see page 15) remove the hex screw that secures the handle.
- Remove the handle and spout.

Removing Washers

- With the handle and spout removed, the parts controlling the water mix will be visible.
- Each brand faucet is different; some will lift out while others require screws or retainer clips to be removed.
- Since there isn't one type of single-handle faucet, an online search by brand will show specific installation instructions.
- Take the part(s) to a hardware or plumbing supply store for the correct replacement.

Replacing a washer isn't especially complicated, but knowing which washer or faucet repair kit to get can be. No two are the same, and not every supplier carries the one you need. A phone call ahead of time with the faucet brand can save you a trip. In the meanwhile, if it's a sink drip and it's really bad, shut the water off under the sink using the shut-off valve (there should be a pair for every sink).

MAKE IT EASY

Shut-off valves are installed at every toilet and sink in a modern home for convenience and safety. They allow you to shut water off for repairs and in the event of a leak without shutting all the water off in the house. If you don't have them, consider having a plumber install them.

Reinstalling the Faucet

- Wipe the faucet body clean and dry before installing the new parts.
- Insert the new parts according to their installation instructions—the packet normally includes a special wrench if needed for the installation.
- It's easy to overtighten plumbing parts and cause new leaks, so carefully hand-tighten any parts requiring tightening.
- Replace the spout and tighten the hex screw that secures it to the faucet, turn the water back on, and check for leaks.

Fixing the Sprayer

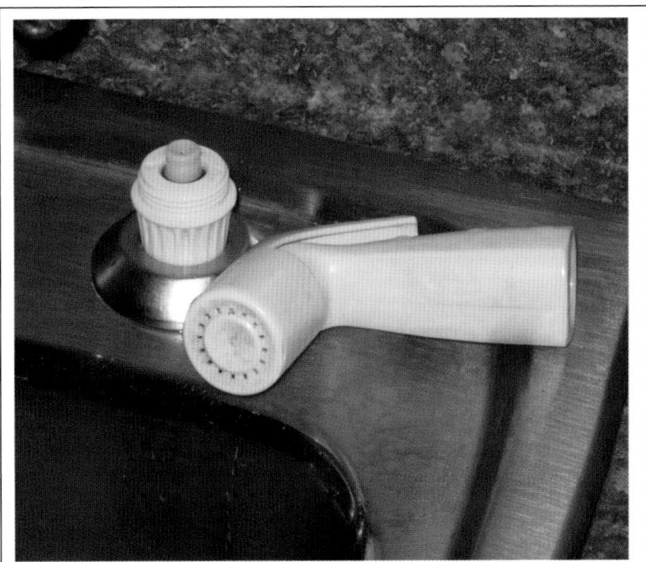

- For a broken sprayer head, shut the water off, unscrew the head, and install a universal replacement head.
- To replace the entire unit, shut the water off, unscrew the hose from the base of the faucet, drain the hose into a small bucket, unscrew the mounting nut that secures the sprayer to the sink, and pull the hose out.
- Install the new sprayer according to its manufacturer's instructions.

TOILETS

A leaking toilet can stem from several problems and should be addressed quickly

A leaking toilet isn't fun. Floors can rot, often unnoticed, from ongoing leaks where the toilet sits on the floor. Water looks for the path of least resistance, and if that's a loose gasket, worn seal, or valve in need of adjustment, that's where it will leak.

Some leaks aren't leaks at all. In very humid climates, a toilet tank can "sweat" due to condensation on the cold exterior and drip water on the floor. If the dripping is excessive enough, it's believed to be an actual leak. Insulation kits for the inside of the tank and special jackets for the outside should resolve this problem. Easier yet, place some plants on the floor under the tank and let them benefit from the dripping.

Running Toilet

- When a toilet is flushed, a rubber flapper or flush valve lifts a rubber flapper or flush valve, and the water passes from the tank into the bowl; as the water level drops, the flapper drops back over its opening.
- If the flapper doesn't completely close and form a tight seal, water continues to run into the bowl.
- When the float, a ball connected by a metal rod to a fill valve, isn't adjusted properly, water continues to run.

Toilet Leaks

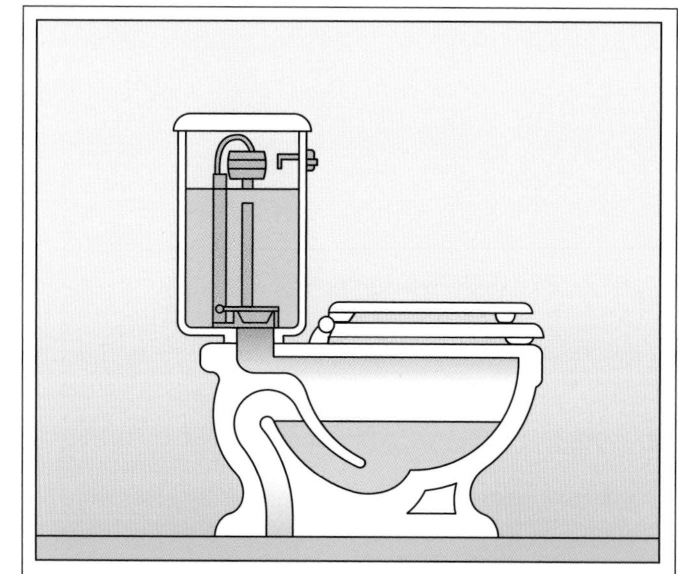

- The most visible toilet leak is at the base of the bowl on the floor and should be addressed immediately before the flooring is damaged.
- A leak between the tank and the bowl indicates a leak in the seal between these two components or at the bolts that connect the tank to the bowl.
- Continuously flowing water is due to a bad flapper valve.
- If water drips under the shut-off valve supplying water to the toilet, it needs to be tightened or replaced.

Leaks from inside the tank or the cold water supply line can be addressed by shutting off the water to the toilet, draining the tank, and repairing the specific components. Leaks at the base of the toilet call for removing the toilet and examining the floor for soft or even rotted wood, which would need to be repaired before reinstalling. With close examination, you can determine the problem and the solution.

YELLOW LIGHT

Never stand on a toilet to get at something overhead. Use a small stepladder or utility stool. Toilets are made of vitreous china and are bolted to the floor as tight as china can safely be secured. Standing on a toilet is inherently unstable and can jar it loose and cause a leak.

The Float

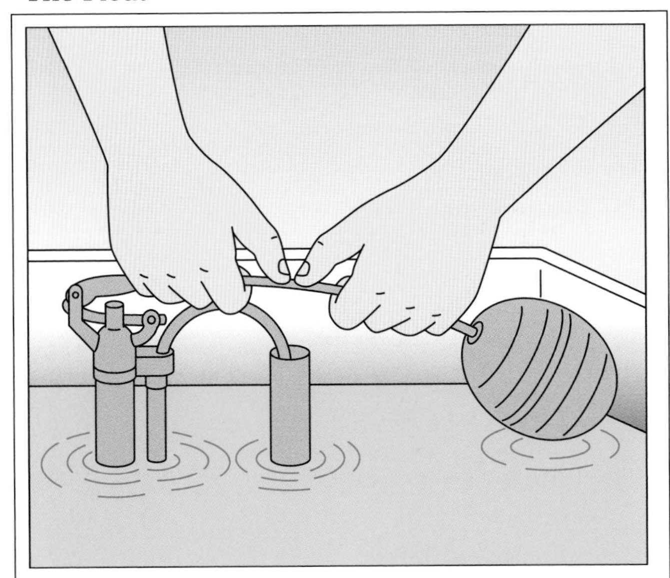

- The float moves towards the bottom of the tank after each flush.
- As the tank refills, the float moves up and should stop when the tank is full again.
- If the water continues to run, pull up on the float until the water stops and then adjust the float to this level using the adjusting screw(s) at the top of the valve.
- Modern-style floats don't use a metal arm, but rather travel up and down the fill valve.

The Flush Valve

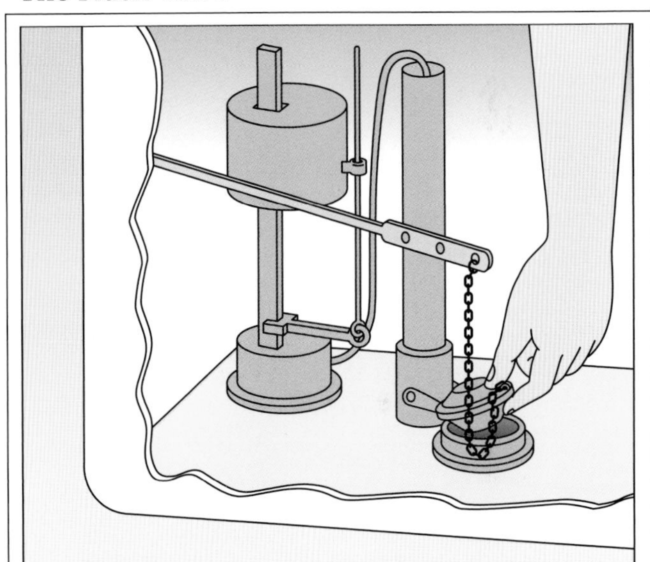

- Check that the flapper chain isn't jammed or the wrong length and caught under the flapper or not allowing it to close completely.
- To replace, shut off the water to the toilet, flush, remove the flapper, and take it to a hardware store for the correct replacement.
- Flappers do not require tools for removal or installation; they just slip over the overflow pipe or are attached with plastic hooks.
- Refill the tank, test the flapper, and adjust as needed.

WASHING MACHINES

Repair washing machine leaks early to avoid big and expensive messes later

We don't give washing machines a lot of thought, but they can be a source of major leaks. Generally, they're dependable and keep running for years, but a washing machine can leak when one of the water supply hoses is damaged or when the machine itself is running. The latter isn't as critical since the amount of water leaked is often somewhat limited. When a hose leaks, though, especially if it bursts, it's a very big deal. Up to 500 gallons of water an hour can pass through a burst washing machine hose.

Leaks while the washing machine is running can result from too much soap, which causes an overflow of suds. A bad tub seal or basket gasket will allow water to leak out as well. A

Shutting Off the Water

- Washing machine water shut-offs are available in various styles, such as a typical hose bib (what you would have outside your house for a garden hose).
- Some older valves that haven't been turned in years can be stiff and difficult to move.
- Newer homes contain shut-offs in recessed plastic boxes inside the wall behind the washing machine for convenience and a finished appearance.
- Various automatic shut-off valves are available, which electronically sense water leakage—especially important if a hose bursts—and shut the water down to minimize damages.

Washing Machine Hoses

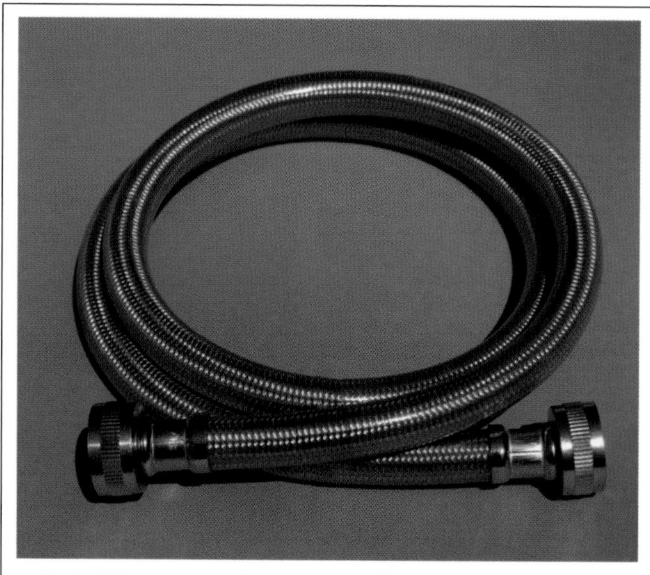

- Typical, low-cost washing-machine hoses are made from unreinforced rubber, although tougher reinforced rubber is available.
- Stainless steel braided hoses are burst-resistant, but no hose is burst-proof; insurance companies recommend replacing any hose within three to five years of installation.
- Allow at least 4 inches of space between the back of the washing machine and the faucet connection to avoid kinking the hose (hoses with right-angle connectors are available if this space is tight).
- Any leaking hose should be immediately replaced.

machine might also leak slightly during one load of laundry and then not again. If the leak is minor, run another load of wash to compare and look for additional leaks. Continuing small leaks should be repaired before they become big leaks. However, if the machine is old and has seen better days, it might be better to buy a new one altogether.

MAKE IT EASY

To avoid damage from leaks, run washing machines (and dishwashers) only when you're home and can shut the water off if needed. Going on vacation? Turn the water off at the shut-off valves or even all the water to the house if it's going to be empty, particularly during cold weather.

Hoses Installed

- With the water shut off, turn on the washing machine to expel any remaining water in the hoses.
- Place a bucket on the floor and loosen the old hoses with a vise grip or channel locks while holding onto the faucet.
- Drain any remaining water in the hoses into the bucket.
- Install the new hoses by twisting the brass connectors, not the hoses themselves; turn the water on, and check for leaks.

Do You Repair or Replace the Machine?

- Some washing machine problems are easily solved: checking the water temperature mix at the faucets, setting the water level per load accurately, being sure the machine is level, and not overloading.
- When a machine suddenly stops, check the power supply first.
- There can be numerous sources of problems: switches, a water pump, motor, bearings, hoses, and electronics.
- A basic washing machine starts at about $300—consider this when major repairs are needed on your machine.

UNCLOGGING SINKS

Unstopping a clogged sink is usually easier than it looks

Why do sinks clog? In the kitchen, any grease, oil, or even soap can stick to the sides of pipes and accumulate over time, grabbing other gunk as it flows by. More often, it's too many food scraps, even when you have a food disposer (not all disposers are created equal). The trap under the sink, a U-shaped section of the drain pipe designed to hold water so sewer gas doesn't seep into the room, is a swell place for food waste to accumulate if it isn't flushed out properly. Empty a sink full of hot water into the drains once a week to help prevent this build-up. Avoid washing food scraps and grease down the drain as well. Collecting grease in a can and tossing it in the trash prevents clogs in your drains and in the public sewer lines, too.

Treating and Preventing Clogs

- Unless a clog is severe and a long time accumulating, it will be located inside the trap and easily removed.
- Chemical drain cleaners are available, but they don't always work as effectively as a basic plunger and can be hazardous if used incorrectly.
- Kitchen drains are generally flushed out more than bathroom sinks with high volumes of hot water from dishwashing, which helps keep them clear.
- Keeping grease and fats out of kitchen drains goes a long way towards avoiding clogs.

Using a Plunger

- Add enough water to the sink to submerge the plunger cup and keep air out of it.
- Cover up the sink overflow drain with a towel; do the same to the clear drain in a double sink when the other drain is clogged.
- Sealing the overflow prevents the plunger from forcing water through it and losing its effectiveness.
- Push down on the plunger for 15 seconds or until the clog is clear. After running some water in the sink, plunge it again.

In bathrooms, it's soap and hair that combine as the number one clogging culprit. For a quick fix, a simple hair screen in both the sink and the tub will reduce the chance of clogging the drain lines.

For easy clean-up, place a bucket under the trap to catch water as you remove it.

MAKE IT EASY

To keep drains clear, pour half a box of baking soda down each drain once a month followed by a cup of white vinegar. The foaming action helps keep drains clean. Follow up by filling each sink with hot water and then opening the drain. The volume and weight of the water also clear the drain lines.

Removing the Trap

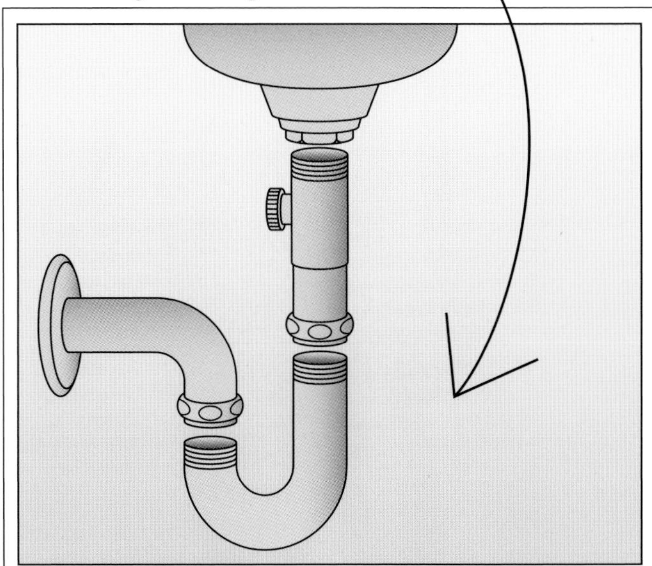

- Every sink has a trap, but there are several different styles and materials.
- Most new installations are PVC plastic and can be disassembled without tools, while older chromed metal requires a channel locks.
- Place a bucket under the trap to catch the water in it and undo the nuts that secure it to the wall pipe and the sink drain.
- Clean out the gunk, wash the trap in soap and water, reinstall, and run some water to test for leaks.

Using a Snake

- Snakes, or augers, come in all sizes from very small hand-operated models to motorized commercial snakes for cleaning out sewer lines.
- Although you can remove a drain stopper and snake down through a trap, you're better off removing the trap first and snaking beyond that if the trap isn't the problem.
- Slowly feed the snake cable into the drain line beyond the trap until you hit the blockage.
- Turn the crank to rotate the auger bit and break up the clog.

UNCLOGGING TOILETS

Fixing your clogged toilet problems, no matter what the source

We stop taking a toilet for granted when it clogs. We want to take care of it fast, get rid of the mess, and return to our daily routines, which, of course, include working toilets.

One source of a toilet clog is obvious to all of us: too much bathroom tissue. Another is tossing in oversized sanitary or personal items that should go in the trash. Items such as small bottles or tubes stored on top of the toilet tank that drop into a toilet as it's going through a flushing action are another terrific way to clog it up.

Young children, alas, find toilets very intriguing. Potty training teaches them to flush, and they begin to wonder what else can disappear down this miniature watery vortex.

Fixing and Preventing Clogged Toilets

- Toilets are designed to move a certain volume of waste and water—anything beyond that and they can clog.
- Tree roots penetrating the drain line going to the street can also cause toilets to back up.
- Never flush a clogged toilet expecting to clear the clog—you'll only fill the bowl to the top or perhaps overflow it.
- As a precaution, shut the water off to the toilet.

Using a Plunger

- A heavier duty plunger designed for toilets with a cone-shaped section that unfolds from the cup is better to use than a standard suction cup style.
- Place the plunger firmly on the bottom of the toilet and plunge vigorously 3–4 times without pulling the cup away from the toilet.
- After the last push, pull the plunger away to allow suction to help break up the clog.
- Repeat if necessary and once the clog is cleared, flush the toilet again.

In goes a stuffed animal, which indeed disappears, but it doesn't get very far. A plunger job now requires the toilet be removed and the line possibly snaked out. Even if your kids unintentionally caused the problem, having them help to fix it can be a valuable lesson for them.

MAKE IT EASY

Leave a plunger in every bathroom. It makes it more convenient when needed and less embarrassing for people who discover they need it, especially a guest who hardly wants to request a plunger from a host to unclog a toilet after using it. Decorative holders are available for storing each plunger.

Using a Toilet Snake

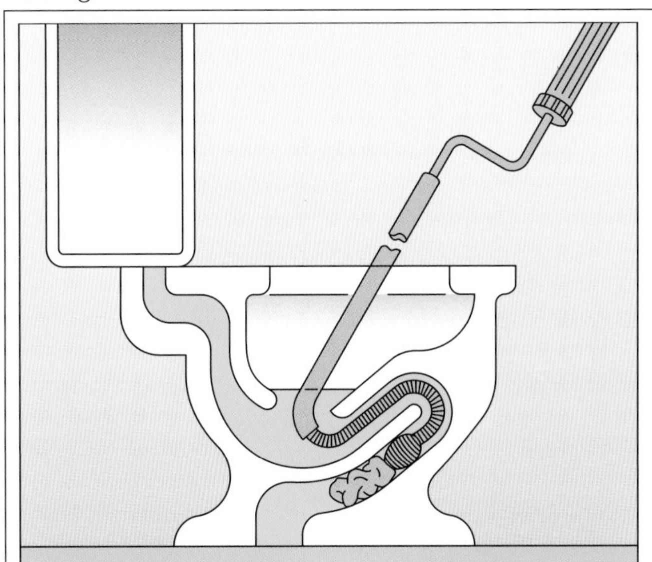

- When plunging doesn't clear the blockage, try a drain or closet auger.
- A closet auger has a fixed length of coiled steel attached to a curved, rigid pipe for accessing the inside of a toilet trap.
- Once the coiled end is fed inside the toilet, turn the handle on the other end and move the snake until it hits the clog.
- Draw the auger back to pull the clogged material into the bowl instead of farther into the drain line.

Fixing Extreme Clogs

- When snaking a clogged toilet fails to clear a local blockage, the toilet will have to be removed. You might want to call a plumber.
- Use a snake that can clean out a 3-inch waste line effectively—try a tool rental supplier.
- If you believe you've cleared out the clog, carefully pour some water down the drain line and when you're satisfied it's getting through, reinstall the toilet.
- Flush the toilet to test the drain line.

UNCLOGGING TUBS AND SHOWERS

A little prevention against build-up keeps drains free flowing longer

Showers and tubs are common spots for clogs and build-ups. You lose up to one hundred hairs a day, and some of them end up in the tub and shower drains. Combined with soap and shampoo suds, this can add up to an eventual clog. How will you know? The tub will drain progressively slower, even when you are showering. You'll find yourself standing in 2 inches of water that shouldn't be there.

Sometimes all it takes to clean out a shower drain is to remove the strainer from over the drain and pull out the matted hair that's slowing the draining. Other times, for more stubborn clogs, it takes a plunger or a plumber's snake, both of which are relatively easy to use. Tub drains are a little more

Unclogging Drains

- Keeping hair out of a tub's drain line is easier than cleaning it out later.
- A simple plastic or metal strainer fitted over the drain opening is great preventative maintenance at almost no cost.
- Pour a kettle of boiling water down tub and shower drains once a week to help keep them clear.
- As soon as water starts to slowly drain, you have a clog and should get to it before it gets even messier to clean out.

Plunging a Shower Drain

- Shower drains are easier to access than tub drains; to clean a shower drain, simply pry off the metal strainer covering the drain line and pull out any hair with a stiff, bent wire (wear latex gloves; this will be messy).
- Cover the open drain with a plunger and push and pull a few times to force any remaining obstructions down the drain line.
- For stubborn clogs, use a small plumber's snake to remove the clogs.
- When finished, pour boiling water down the drain.

involved. There is some disassembly required, some cleaning, plunging, and perhaps some snaking. In the future, a little prevention goes a long way to avoiding this messy job again.

GREEN LIGHT

It is a good idea to make a habit of cleaning out your drain or emptying the trap after every shower or bath.

Use needle-nose pliers to clean out an obstruction in a tub drain.

Cleaning Out a Tub Drain

- Drains without trip levers —which are found on the tub wall near the faucet— have stoppers that screw directly to the drain pipe.
- To remove, turn the stopper counterclockwise until it's loose and pull it from the drain.
- Clean out all obstructions—a needle-nose pliers is a useful tool for this job—and clean the drain opening and stopper with disinfectant cleaner before reinstalling.
- Use a plunger or snake for stubborn clogs and pour boiling water down the drain.

Tubs with Levers

- Tub drains with a trip lever can be trickier to disassemble and clean.
- A trip lever is connected to a lifting rod, which moves a plunger or a drain stopper that opens and closes the drain.
- Loosen the screws on either side of the trip lever and pull the mechanism up and out of the tub.
- Remove the drain opening or drain stopper, depending on which type of drain you have, and clean the drain out.

ROOFS

Fixes for patching a leaking roof and knowing when to call in a professional

Roofs leak at the most inconvenient times, such as when it's storming outside, when we would prefer the rain also stay outside. Diligence and yearly inspections will head off some leaks but not all. Check for aging roof shingles and deteriorating flashing, which, when combined with bad weather, can cause leaks. On older roofs, tell-tale signs will be areas that have been patched with black roofing tar, which often indicates a previously patched leak. These areas are more vulnerable to future leaks.

One problem is finding the actual source of a leak. Water leaking through a roof then follows the easiest pathway along attic rafters and framing, which can be several

Attic Leaks

- Unfinished attics provide a better chance of finding the source of a roof leak than finished attics (you'll probably have to remove some wet ceiling drywall in a finished space).
- When it's raining, find the leak and follow it back to the underside of the roof.
- If it's a direct overhead drip, consider yourself fortunate; your job is easier.
- When the water runs along rafters and supports, follow it carefully with a flashlight, checking that you've found the accurate spot.

Driving a Nail

- Once you're sure you've found the source of the leak, drive a framing nail up into the roof from inside the attic.
- Note approximately where the nail is located so you'll know where to find it on the outside of the roof.
- The nail won't make the leak any worse but allows you to locate it.
- On the roof, pound the nail back in and patch its hole plus the original leak.

feet away from the leak's point of origin. Even experienced roofers can be puzzled at the source of a leak. This is one reason why some of those tar patches have been applied more than once and the leak still doesn't stop—even if the patch appears to be in the right place. Until the leak can be identified and sealed, a bucket will be your best stop-gap measure.

YELLOW LIGHT

Wear nonslip shoes and take your time when walking on your roof. Secure your ladder by pounding stakes in front of its feet if the ladder is on soft ground or by tying off the bottom rung to a secure point on the house if the ladder is on concrete.

Containing the Water

- Place a bucket under the leak and observe how fast it fills up.
- Chances are, the leak will be slow and can even be ignored for the moment as long as it isn't soaking insulation or dripping near wiring.
- If the bucket fills slowly, the water will evaporate during dry days.
- After the leak is repaired, keep the bucket in place for a season and check it for water after rainy days, especially rainy days of high wind.

Roofing Cement

- Roofing cement is a universal quick fix for leaky roofs but is not a permanent repair.
- Many leaks occur at the metal flashing, which acts as a transition from the roofing shingles to chimneys, roof valleys, any protrusion through the roof, and any house walls.
- Use sections of metal flashing to patch holes in existing flashing or to replace missing shingles when no spares are available.
- Slide the flashing patches up and under the shingle(s) above its installation point.

GUTTERS & DOWNSPOUTS

Stop leaking gutters and avoid bigger problems like wet foundations later

Most houses have rain gutters. The gutters' main purpose is to funnel and direct rain away from the overhanging sections of the roof, the siding, and the foundation. Modern gutter systems empty into storm drains through the downspouts. Others have downspouts that empty onto splash blocks and allow water to empty into the yard, yet still away from the foundation. In wet climates, the farther you can keep water away from the foundation, the better.

Gutters were once made from wood. Some of these still remain on older homes, and replacement wood gutters are still manufactured, although they are expensive. With care, they will last for decades. Copper is the next-costliest mate-

Gutter Leak

- When gutters are installed with the proper slope to avoid standing water, they have a better chance of avoiding holes and leaks.
- Over time, steel gutters can rust and wood gutters can develop holes, but copper gutters will not rust.
- Water also accumulates when gutters are not cleaned regularly.
- Any place two gutter sections are joined or a downspout is installed there's a potential for leaks.

Gutter Seams

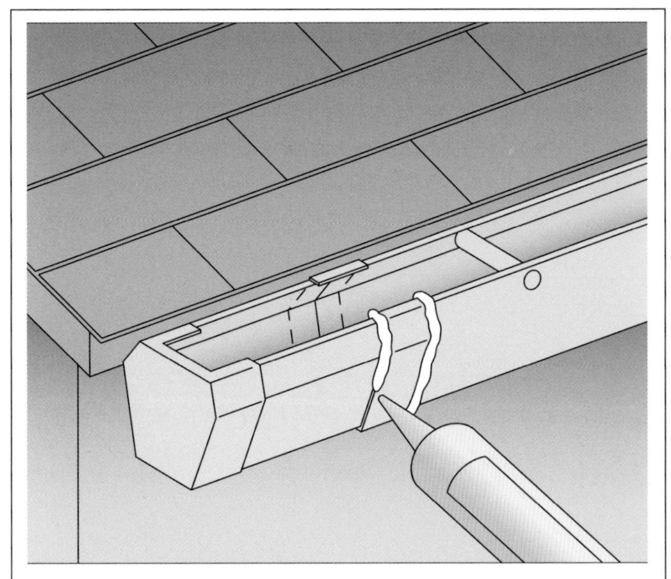

- Aluminum gutters are sized and cut to fit, but individual sections are joined at corners forming seams.
- The seams are sealed with gutter and flashing caulk, which is specially formulated for this application although silicone caulk will work if necessary.
- Some leaks show up in wind-driven rain only, and the resulting damage takes years to show up.
- Reapplying a generous amount of caulk to any suspect seam should stop any leaks.

rial and rarely seen in new homes except on very high-end construction. Steel gutters have mostly been replaced with aluminum, the most popular choice in modern gutters. Vinyl gutters are more a do-it-yourself choice. Each material has its advantages and drawbacks, and all types, with maintenance and care, will give long-lasting service. Leaking gutters, like any source of leaks, should be repaired before they cause greater damage elsewhere. Water goes looking for trouble whenever it gets the opportunity. Plug the hole, seal the gap, and keep water flowing where you want it to.

ZOOM

Roof-mounted rain water collection systems and troughs have existed for thousands of years. In England, the disbanding of the monasteries in 1539 made large quantities of recycled lead available for elaborate gutter installations. Colonial-era gutters in America were initially made from wood and later lead and eventually other metals including copper and tinplate.

Epoxy Repair

- Repair a leaking wood gutter with waterproof epoxy. Clean out the rotted material from the dry gutter, brush it with a wood preservative, and when that cures, fill it with the epoxy according to the manufacturer's instructions.
- Repair rust spots in metal gutters with a wire brush, spray it with rust preventative paint, use an automotive fiberglass kit with mesh screen to repair, and spray paint the patch.
- Any patch should be monitored and considered impermanent.

Testing with Water

- After your repair cures, fill the gutter with water and check for additional leaks.
- If the water is not draining properly towards the downspouts, the slope of the gutter needs adjustment.
- Check the gutter hangers for tight fasteners and bends from supporting too much weight.
- Adjusting the hangers, bending them, or even adding to them can readjust the slope so the water drains correctly.

WINDOWS

Seal window leaks before they seep into walls and cause more damage

Old windows can go decades and never leak. New windows can leak during the first good storm. How come? It has to do with the installation and the flashing. A properly installed window with the correct flashing and caulking should stay leak free indefinitely. A bad design or construction detail can cause even the most expensive window to leak. The leak most often occurs where it's attached to the framing or when water gets behind the flashing and then behind the window. Improperly flashed windows are a regular cause of water infiltration in homes.

Old windows can leak at the glass when the glazing compound or putty deteriorates and falls off, which always hap-

Windows

- Windows leak at the glass, the glass seals, and the flashing.
- Older wood windows can leak where the glass butts up against the wood sash if the exterior glazing compound is missing or badly cracked.
- Caulking should not be used to take the place of proper flashing but can supplement it.
- Rare leaks from unusually hard driven rain should be monitored but are probably unimportant.

Cover Window with Plastic

- Whenever a window leak is elusive or if excess water comes in, cover the outside of the window with plastic to prevent further damage.
- If water still comes in, the leak could be coming from higher up in the wall and dripping down through the window.
- Wood windows are particularly susceptible to water damage, which can lead to rot.
- Tack the plastic up using wood strips and small nails or exterior-grade duct tape.

pens if the putty goes unpainted and unsealed. They can also leak during periods of wind-driven rain that blows in from the window sill. Regardless of whether it's old or new, the main problem with leaking windows is water seeping into the walls and causing damage. Although it sounds like a broken recording by now, seal these leaks before they cause you bigger headaches.

Interior Moisture on Single-Pane Windows

- Older wood and metal windows are single pane—one pane of glass versus new insulated windows with two or three panes sandwiched together.
- In cold winter climates, interior moisture condenses on these panes and can give the appearance of leaks.
- Excess condensation will corrode steel windows and deteriorate wood windows, so routinely check the windows.

MAKE IT EASY

Not sure where a window is leaking? Cover it with plastic until it dries out and the weather clears and then examine it thoroughly on the outside for cracks, loose caulking, or separating. It's important you stop the damage, and plastic does the job. Keep a roll of 6mm clear plastic around for this and other jobs.

Caulking

- Caulking is used to prevent water from washing through seams in the exterior window trim or where the trim and sill meet the wall of the house.
- Caulk isn't a miracle cure, and overly thick applications can fail.
- Old, worn caulk can be removed with a utility knife and replaced before applying new caulk.
- Be sure to caulk during dry conditions only.

UNCLOGGING GUTTERS

Tackle this problem to avoid damage to house paint and siding

Gutters are intended to collect water running off your roof, but they also catch leaves, twigs, granules from roof shingles, and tennis balls tossed on the roof. Some of this debris will wash down (and possibly clog the downspout), but after enough of it accumulates, it will slow down the flow of water to the point the gutter will overflow. An overflowing gutter means trouble because sheets of water flowing down the house siding will lift entire sections of paint. Eventually, the wood will deteriorate or, worse, leak inside your house.

There are new gutter cover designs that claim you'll never have to clean your gutters again. There are many of these products, and some do a decent job of keeping large leaves

Clogged Gutter

- Gutters get clogged with anything that can fall from a tree or blow up on a roof.
- As clutter accumulates, it holds water, which adds to the gutters' weight.
- Heavier gutters can pull away from the roof, both loosening them and affecting their slope, adding to the drainage problem caused by the debris.
- There is no recommended number of cleanings throughout the year, but do a minimum routine fall and spring cleaning.

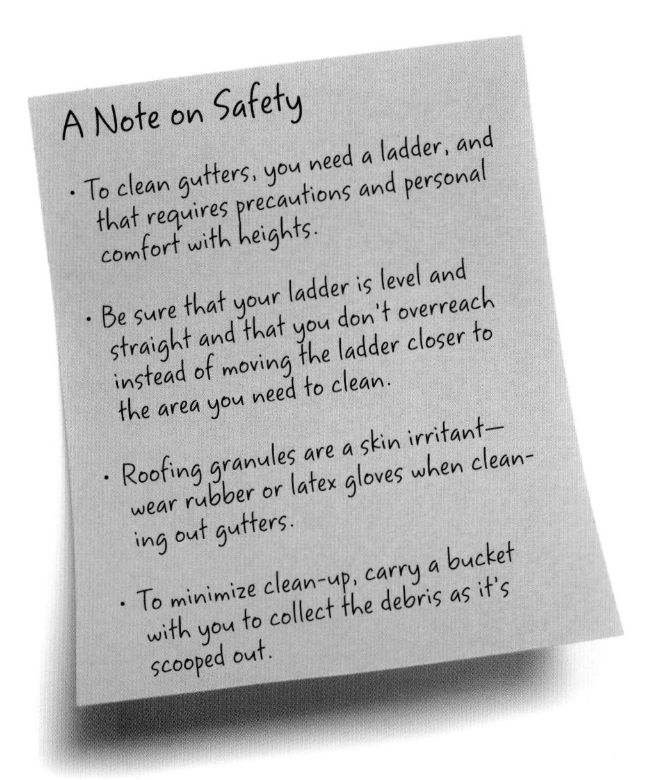

and twigs out but not smaller seeds, pods, and dirt. There will probably never be a gutter that doesn't need some cleaning, and unless a gutter cover is easy to remove, these new cover designs might actually make gutter cleaning tougher to do.

Regular cleaning prevents clogging and provides an opportunity to do some roof inspection while scooping out dead vegetation. How often you clean gutters depends on how full they get. Seasonal checks will let you know.

YELLOW LIGHT

If you use an extension ladder, secure the legs to prevent slippage. A stepladder should be level and opened in a fully extended and locked position. You can't grab onto a gutter and depend on it to support you. Your ladder should be sturdy enough to allow you to safely do the job.

Gutter Scoop

- There are tools purported to let you easily clean lower-level gutters from the ground, including tongs, water blast systems, and scrapers.
- Unfortunately, cleaning gutters requires you see what you're doing up close with a lot of hand scooping if the debris is wet.
- If the debris is dry, a lot of it can be removed with a blower.
- A bleach bottle can be cut into a scoop shape for an inexpensive gutter-cleaning tool.

Hosing the Gutters

- Don't try to wash a large volume of gutter debris out with a garden hose, as it will push too much of it into the downspouts, where it can cause more clogs and fill the gutter with water.
- After scooping out the gutters, hose down the gutters.
- To avoid clogging the downspouts, loosen and remove their bottom ends to keep any debris out of the drain line.

UNCLOGGING DOWNSPOUTS

Clean out downspouts at the same time as your gutters to avoid water damage to the house

A clog in a downspout is not as easy to spot as a clog in a gutter. The clog could be at the top, in the elbow, or near the bottom of the downspout. Sometimes a hosing will clear it out, and other times this just forces the clogged material farther into the downspout or drainage system.

Some downspouts are made of steel, and the elbows, the curved sections that often connect the gutters and the downspouts, can rust if they get too full of debris. The debris clogs the elbow, which causes water to build up here instead of passing through and allowing the elbow to dry out.

Tips for Cleaning Downspouts

- Before cleaning the gutters or the downspouts, pull the bottom ends of the downspouts out and away from the drain lines.
- Downspouts usually get clogged in the upper bend and can be hosed out.
- After cleaning the gutters, hose them out and check for blockage in the downspouts.
- Older downspouts are narrower than the gutters and can easily get clogged with leaves and debris.

Hosing Out the Downspout

- Pull out the clogged material from the bend with a piece of bent wire with a hook on the end before snaking the bend.
- The bend can be corroded on the inside due to years of standing water, so be careful with the wire, particularly with steel downspouts.
- Insert the hose nozzle at the top of the downspout, wrapping a rag around it for a tight seal.
- Open the tap full force until the water shoots out at the bottom of the downspout.

To fix this, you might have to dismantle the downspout. Some downspouts are easier to loosen and remove than others should this be necessary to clean them out.

To prevent this mess, downspout screens are available to keep leaves and large debris out. Be sure to routinely clean the area around the screens to avoid clogs at the top of the downspout.

GREEN LIGHT

Keeping your gutters and downspouts clear will help prevent future problems. See the maintenance timeline on page 216 for friendly reminders on when to do things throughout the year.

Snaking the Downspout

- If a snake (see page 67) is necessary, use a manual one and turn it slowly, pulling and pushing in and out until the clog is broken up.
- Run the hose into the downspout to check that it's clear.
- Run the hose inside the drain line as well to confirm it's clear.
- Reconnect the downspout to the drain line and check that it's secure at the gutter as well.

Preventing Downspout Build-Up

- Keep leaves and twigs out of the downspouts by installing a leaf guard over the top of each downspout.
- Remove accumulated leaves before they cause a water back-up.
- Avoid leaf guard systems and screens for entire gutters, as they often do not prevent smaller debris from accumulating in the gutters and can make cleaning difficult.
- Add extensions onto downspouts that empty onto splash blocks instead of drain lines to assure the water empties away from the house foundation.

UNCLOGGING OUTSIDE DRAINS

Fixing outdoor drains prevents floods and allows them to do their job properly

Modern homes often have driveway drains and yard drains. Some of these are connected to the local storm sewers, and others empty into French drains in the yard itself. A French drain, named after Henry French, is essentially a trench or pit filled with gravel or other small stones, constructed to absorb and reroute water. The top is usually covered with grass to hide it. The thickness of the pipe influences where it can be used in the yard and how susceptible it is to damage.

Older drain lines consist of sections of clay or concrete pipes butted together and often are damaged with cracks and intrusive tree roots. In the event of a clog, there is only

Good Things to Know about Yard Drains

- Tree roots will press against any type of drain line and infiltrate openings or cracks looking for water.
- Expert drain-cleaning services clean out and cut tree roots with special snakes, but the roots can return.
- To clearly view and determine the cause and location of clogged sewer and drain lines, some services inspect the lines with a miniature camera.
- Consult local regulations to determine who pays for cleaning out your drain lines if a neighbor's tree causes a clogging problem.

Driveway Drain

- Driveway and yard drains, as well as any foundation drains, typically empty into your city's storm sewers.
- They remove excess rain water and snow melt, preventing overly soggy yard areas that could lead to problem soil conditions.
- For yards lacking drains but needing to control runoff or stop water accumulation, French drains—trenches full of gravel or perforated drain pipe—are an option.
- Adding drainage can affect a neighbor's property and might require a permit from your building department.

limited cleaning out you can do by yourself, and some drains need to be replaced altogether.

How do you know if a drain needs to be replaced? Drain cleaning companies and plumbers send special cameras down through the pipe to determine exactly what's causing the blockage. This is the only way, other than digging, that you'll know whether your drain lines are repairable or not. A camera probe is much cheaper than replacement, so enlist the aid of a professional before you decide on replacement.

YELLOW LIGHT

Be careful using drain rooters when cleaning out plastic drain lines. The force of spinning blades can destroy perforated, thin-walled drain lines, which can be cleaned by hand-driven snakes or high-speed water jets. If you're unsure what your drain lines are constructed from, use a conservative approach when cleaning them.

Hosing the Drain Line

- Yard drains receive a lot of dirt and silt, which gradually accumulate in the drain until it reaches the drain line.
- Check your drains every few months and scoop out the dirt before it becomes a problem.
- After removing the dirt, run a hose down the drain line at full force to confirm there isn't any blockage.
- Keep the drain covers clear of leaves and grass clippings whenever you're doing yard work.

Snaking the Drain Line

- Drain lines in your yard can get clogged and require cleaning to avoid overflowing.
- Only hard materials—rigid plastic, tile, and concrete lines—can withstand cleaning and clearing with a powered plumber's snake.
- Corrugated drain lines, thin-walled plastic popular for drainage because of easy installation, are inexpensive but not built to withstand mechanical cleaning.
- Short of digging up corrugated lines, run a hose in with a high-pressure nozzle to eliminate blockages.

FLOORS: WHAT TO TRY FIRST

Squeaky floors are annoying—here are the initial steps to eliminating them

Floors are usually composed of a subfloor—plywood, particleboard, or individual tongue-and-groove boards—and either carpet or a finished wood floor. In some older homes, the finished wood floor is thick enough that a subfloor is not needed. Over time, floors can squeak as nails loosen and boards shift. Gluing down the subfloor helps eliminate squeaks and movement in the floor, but this is not a universal practice during home construction and was not done in older homes.

Moisture adds to the squeaking problem. During seasonal changes, unfinished wood can absorb and later lose moisture. The wood expands and shrinks, the result being the squeaking noises you hear.

Under the Floor

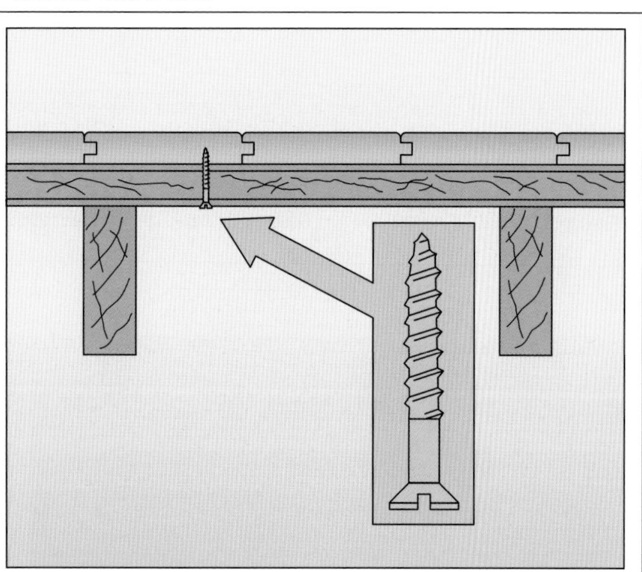

- Finished wood floor boards can rub together, causing squeaks.
- Filling the seams between these boards with powdered graphite talcum powder or spraying with silicone can temporarily stop squeaks.
- Longer-term solutions call for some combination of shims, cross supports, bracing, or fasteners,.
- When space under the floor is exposed, access is provided to the joists and subfloor, allowing reinforcement and "beefing up" of the area around the squeak.

Installing a Shim

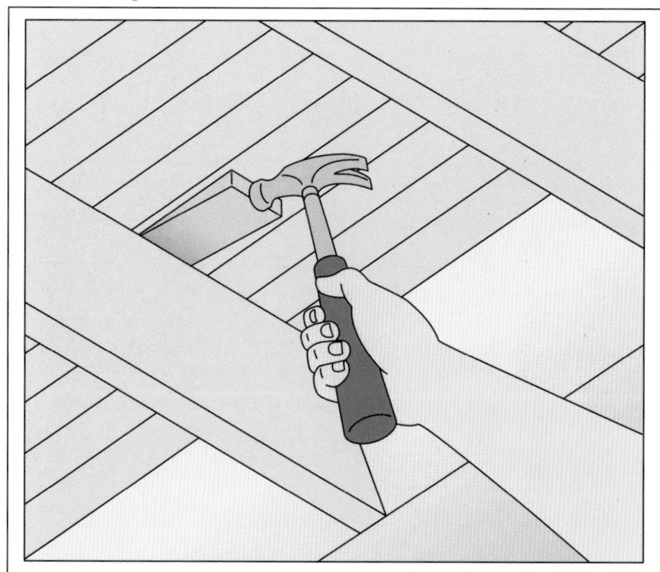

- While someone walks around the squeaky area, go below to the exposed joists until you locate the noise.
- Spread glue on a shim and hammer it between the joists and the subfloor until it's just tight.
- Have your helper walk around further to confirm the squeak is gone or you need more shims.
- A cedar shingle can easily be split into shims for this repair.

You can take on some of the simple solutions below to eliminate or diminish squeaky floors, especially floors whose underside—the joists and the subfloor—is exposed and accessible. It can take some time finding and fixing the problem, but it's better than listening to squeaking every time you walk by.

MAKE IT EASY

If your floor squeaks according to the season, consider sealing the exposed wood—with paint, varnish, or most any finish readily available—as part of your solution. Instead of buying specialized hardware, simple angle irons, which are more readily available, can be screwed into joists and subflooring to eliminate or reduce squeaks as well.

Adding Support

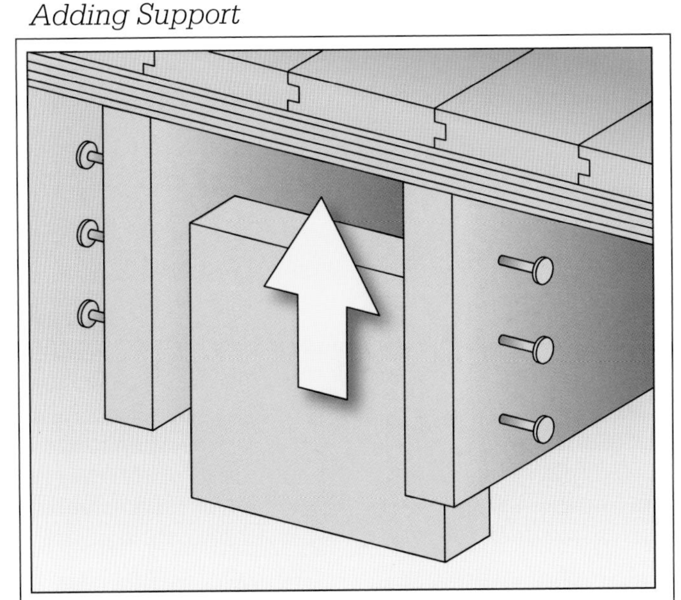

- For larger areas, screw and glue a 2"X2" board into the joist and into the subfloor itself.
- Be sure your screws don't go all the way through the subfloor.
- Screw and glue a section of 3/4-inch plywood—wide enough to reach each joist—into any two adjoining sections of subflooring under a squeak.
- Nail a 2"X6" between the shimmed joist and one or both of the closest joists with the width of the 2"X6" running against the width of the joists.

Hardware

- Screw-in hardware (to tighten loose subfloors and joists) is available from various manufacturers.
- This hardware is especially convenient where there isn't much room to swing a hammer, but there is enough room to use a drill.
- Inexpensive angle irons can be purchased at any hardware store.
- Angle irons offer the option of being available in longer lengths, which can brace a larger area of flooring.

FLOORS: WHAT TO TRY NEXT

With no access underneath the floor, you need a different strategy to fix these squeaks

Squeaky entry-level floors with open basement ceilings are easy to get at and fix, but what do you do with a squeaky second or third floor? You need to approach the repair directly through the top of the floor itself, and this can be intimidating. These repairs are a minor challenge, and you may want to call in the help of an expert for this process. But with care and attention, you can do them without much trouble. The first and main hurdle is to locate the floor joists.

Like floors, stair treads can also squeak since they're used so heavily. The same repair issues and techniques apply. All of these repairs call for the least amount of intru-

Finding the Joists

- Joists support floors and are normally 16 inches apart, running in the direction of a building's narrowest dimension.
- Tap (or use a stud finder) on the ceiling below the joists until you hear a dull thud—this is a joist. Other joists are near it.
- Starting at an outside wall, tap on the floor with the handle of a hammer to find a joist.
- Carpet over plywood/particleboard can be loosened and rolled back to find joists.

Drilling into a Wood Floor

- Determine or make your best guess where the joist is located and follow it to an exterior wall.
- At the edge of the baseboard, use a long, narrow drill bit (7/64 X 6" or so) to drill down through the subfloor and into the joist.
- It might take several test holes to find a joist.
- Once the joist is found, follow it to the area of the squeak.

sion, so use only as many screws or nails as necessary to eliminate the squeak.

Have patience. It might take a few tries to find the joist.

YELLOW LIGHT

As an alternative to a standard screw, try Squeeeeek No More or a similar system that breaks the screw head off after it's inserted. If the carpet has been pulled back and you can countersink the screw, then a standard screw or long deck screw will do.

Installing a Fastener

- Drill one pilot hole per nail or screw into the joist near the squeak.
- Nail in a 16d finish nail, and set it below the top of the finish floor or carpet.
- Insert a 3-inch drywall screw using a drill and Phillips screw bit—if going through carpet, tape the screw threads to prevent them from grabbing and pulling at carpet threads.
- Drive the screw head until it's flush with the subflooring; fill in any wood floor holes with color putty.

Stair Treads

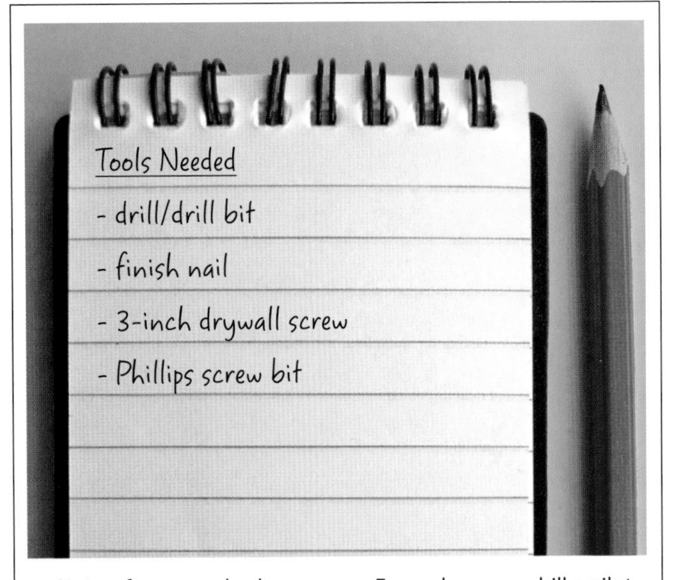

- Stairs often squeak when the tread, the part you walk on, loosens and begins rubbing. The repairs depend on whether the underside of the stairs is exposed or not.
- From an exposed underside, glue and nail a strip of wood a foot long into the riser and the squeaky tread.
- From above, predrill a pilot hole through the nose of the tread into the riser below and secure with nails or screws.
- Toenail (nail at an angle) the tread into the stringers for further tightening.

STICKING DOORS

A sticking door can pull paint off the jamb—fix it and avoid this

Doors can function for years and then start sticking. A slight shifting in the door jamb—the frame that the door is installed in—is one possibility, as is the gradual absorption of just enough moisture to finally expand and no longer fit the jamb as it once did. Any door can expand; it's not limited to exterior entry doors. Any time the door no longer fits and clears the jamb it will stick and bind.

Some select and limited trimming of the sticking surfaces generally resolves the problem, but the key is to trim only as much as needed and not to be too aggressive about it. Remove too much, and you get a gap for hot air out in the winter and cold air in.

Slightly Open Door

- First, examine the door and determine exactly where it's sticking.
- Unless all four edges of a door are sealed, a door can absorb moisture, expand or warp, and stick.
- Shifting in the door jamb or the wall framing can cause a door to rub, but this is very infrequently the problem.
- Loose hinges or a loose lockset causes opening and closing problems, as can excess paint build-up on the edges of the door and on the jamb.

A Quick Fix: Paraffin

- Some sticking is seasonal, and a little paraffin or candle wax can get you by until the weather turns drier.
- Determine where the door is sticking, check that the hinges are tight and the lockset isn't sticking, and rub the edge of the door with paraffin or bar soap.
- After testing the door, wipe off any excess soap or wax with a clean rag.

Doors typically have 1/8-inch constant gap between their edges and the jamb for clearance, ease of opening and closing, and to accommodate weather stripping at exterior doors. Yours might not be this exact, but you don't want it to be any bigger as a result of your repair.

YELLOW LIGHT

If you intend on painting the door, remove all soap or wax from any painted areas before recoating. If after the removal the door sticks, it will need to be trimmed or planed down some before painting. Otherwise, it can pull the new paint off the jamb or door edge.

Trimming the Door

- Mark off the sticking area on the jamb with a pencil.
- Use a very sharp paint scraper to shave all the paint off until exposed wood is visible.
- If the door still sticks, trim off the corresponding area on the door edge and sand any scraped areas with 100- and 120-grit papers until smooth.
- If the top or bottom of a door needs trimming, keep the door in place and use coarse sandpaper instead of a scraper.

Priming and Painting the Door

- New finishes stand out against existing ones, so minimize the amount of recoating while still sealing exposed wood.
- Always paint exposed top and bottom edges of a door—these easily absorb moisture and expand.
- Tape off the interior and exterior areas near the edge so the new paint stays only on the edge.
- Do the same on the jamb.

STICKING WINDOWS

Learn some simple solutions for sticking windows, even new vinyl ones

Sticking windows occur when tracks wear down, paint gums things up, and weather stripping gets slightly bent. Old wood windows are generally the stickiest, and they should be; they've been around longer, have moved more often, and have been painted over and over again to the point that some barely move at all. All windows need some attention, though, regardless of the material they're constructed from or their age.

Wood casement windows open and close like doors, and sticking ones can be repaired the same way as sticking doors (see page 88). Metal sliding windows, whether steel or aluminum, normally don't have paint problems as they typically remain unpainted, but metal-on-metal movement calls for lu-

Vinyl Windows

- Open the window completely and thoroughly clean the track with #000 steel wool and any liquid automotive polish/wax.
- Wipe away any residue and test the window.
- Rub the track with additional wax or spray with silicone.
- Spray with silicone once a month and clean the tracks again as needed (yearly or so), but avoid getting any cleaners or spray lubricants on the brush-style weather stripping often found with vinyl windows.

Wood Double-Hung Windows

- Spray any area where the window sash moves with plenty of spray silicone or WD-40, moving the sash repeatedly until it moves smoothly.
- If the sash is especially stiff, take your time moving it—if movements are too abrupt, the ropes can break.
- If excess paint is the problem, open the sash as far as possible, and scrape and sand any painted edge that the sash slides against.
- Touch up any scraped areas.

brication that these windows can be sorely lacking. Like metal windows, sliding vinyl windows, the latest and greatest version of home windows, are also prone to sticking. Vinyl, like wood, expands and contracts as the temperature changes, another factor in sticking.

Removing surface corrosion, dirt, and grime goes a long way towards alleviating sticking windows. Stripping or sanding the paint off a window is another option for stopping sticky windows long-term. These are homeowner-friendly projects and don't require a professional contractor.

ZOOM

Paint build-up on wood windows can most effectively be removed by disassembling the window, removing the sash, and stripping or sanding the paint off. This is a more-involved task but can be done by a homeowner. See the Resources section on page 228 for more on repairing wood windows.

Wood Casement Windows

- Determine where the window sash is sticking, mark it, and trim down that section of the jamb with a sharp paint scraper.
- If necessary, scrape down the edge of the sash as well, but be careful around any integral weather stripping often found on newer casement windows.
- Test the sash, and continue trimming if needed.
- Sand, prime, and paint all scraped areas, taking care to limit your painting to these areas.

Aluminum Windows

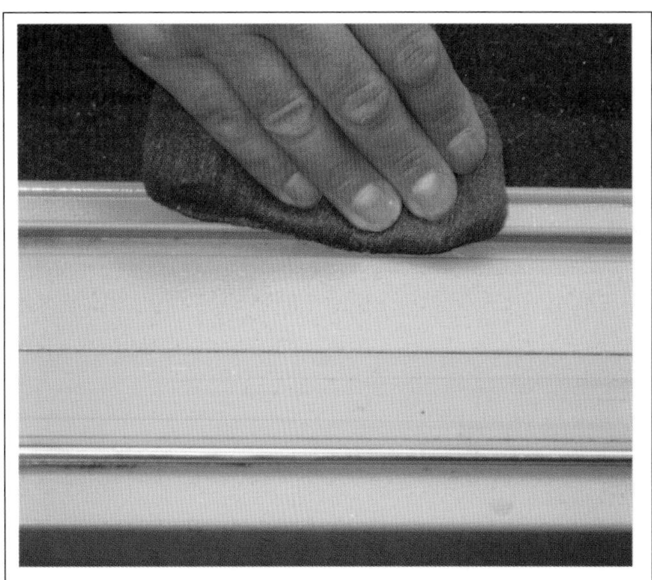

- Sliding aluminum windows last forever and won't warp, but they can bind after years of use.
- Clean the window track with #000 steel wool and naval jelly, a mildly acidic cleaner.
- If naval jelly is not available, automotive polishing compound (see page 46), either liquid or paste, or liquid cleanser will do.
- Wipe off all cleaner residues, apply a coat of automotive wax, and lubricate all locks and hardware.

STICKING FURNITURE DRAWERS

Keep drawers from sticking and make opening them easier

Wood drawers built before the era of drawer slides—those wonderful metal supports that run along the sides of drawers—can swell up and stick shut depending on weather conditions, loose corners, splits in the sides, or a loose bottom.

Repairs are not complicated and do not require special tools. If you can't find the specific problem, just do a general overhaul. It won't take long to sand, seal, and lubricate a single drawer and get it back in action. You'll get more noticeable results with smaller drawers than larger ones, but both will benefit from a tune-up.

Newer drawers with sliders or runners can use some attention from time to time as well as when the sliders begin

Wood Drawer

- Furniture and cabinet drawers stick when they become misaligned after repeated use or they absorb moisture and expand.
- Older drawers slide wood on wood, which can bind.
- Newer drawers move on metal slides whose screws loosen—remove the drawer and tighten these, and apply paraffin or spray silicone to the slides.
- A broken drawer with loose corners or bottom will need repairing before adjusting it for movement.

Wood Drawer without Runners

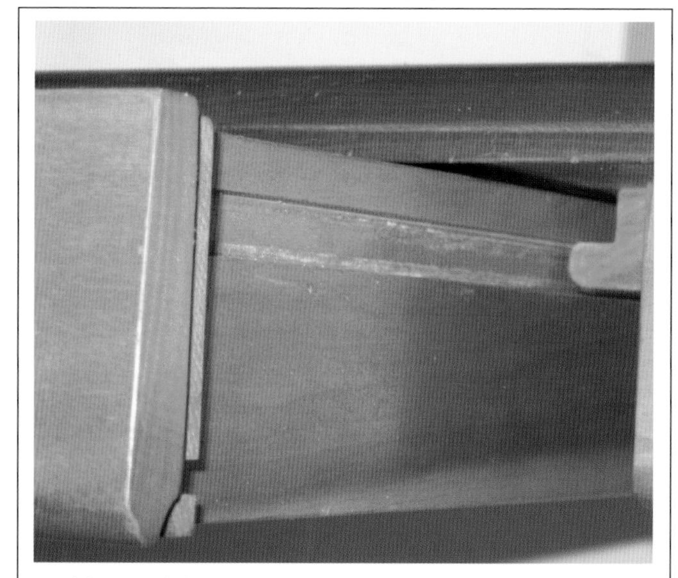

- Older wood drawers slide within an opening using their own sides or small blocks as the guides; these wear down over time and need upkeep.
- Remove the drawer, and rub the sides with paraffin or soap, or spray with silicone; check alignment.
- Lightly sand down tight drawers and then apply paraffin, soap, or silicone spray.
- A wood drawer might need a shim or guide added for smoother movement.

to loosen. Once you know how to repair these drawers, you won't have to yank on your dresser just to get dressed in the morning.

Sealing the sanded areas with an oil or wax will help keep your drawers gliding smoothly.

YELLOW LIGHT

If you have a true antique, consult with a dealer or restorer about how much sanding or alteration you can do to a drawer before you affect the value of the piece. Your treatment options might be limited in order to retain the originality and appearance.

Sanding a Drawer

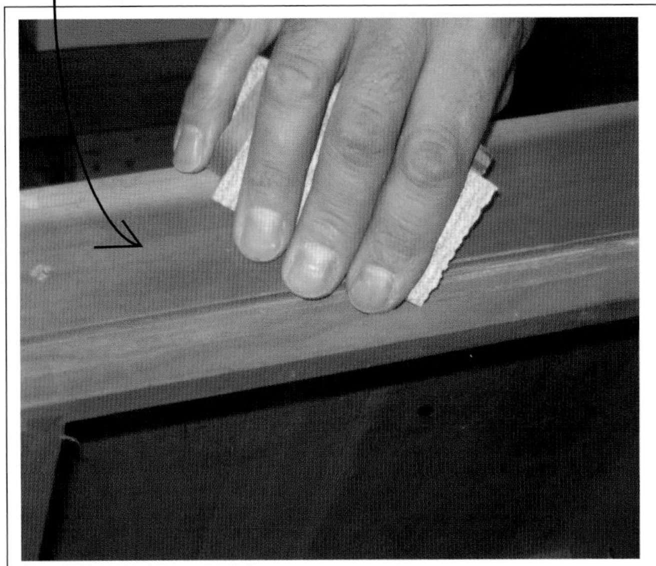

- An electric sander makes short work out of some sticking drawers, especially the fronts if the edges are gummed up with paint.
- Remove the drawer, and sand the sides and edges with a speed block sander and 80-grit paper, smoothing with 100-grit paper (with the drawer out, sand stains on the inside, too).
- Don't worry about taking too much wood off the sides; a speed block isn't that aggressive.
- Seal the sanded areas with a penetrating oil or wax.

Furniture Drawer with Sliders/Runners

- Purists won't install metal drawer runners on some old furniture, but they do make life easier and can prevent wear and tear to drawers.
- Several styles of drawer runners are available with different mounting locations.
- The guides are aligned and screwed to the drawers and its cabinet or the furniture piece containing the drawers.
- No cutting or alterations are required, but start with a small drawer, if available, to get a feel for the installation.

STICKING KEYED DOOR LOCKS

Repair your jammed door lock before it puts you in a jam

Most entry doors now have two locks: one in the door knob and a dead bolt, both keyed the same so, as a convenience, only one key is needed to open them. These locks are typically a pin-and-tumbler design, which features a series of small pins of different lengths. When a key is inserted, the key's notches are cut to match up with the pins and move them to either lock or unlock the door.

Moving metal lock parts, like moving wood parts, can wear down, stick, or even jam up after enough daily usage. Locks can also get stuck in the door strike, the metal section of the lockset installed in the door jamb. The sliding bolt part of the lock slides into the strike, and if the strike is misaligned or not set correctly, the lock will stick and be difficult to operate.

Checking with Key

- Test a sticking door lock by opening the door and moving the lock with the key and the thumb turn.
- If the lock does not move freely, then it's sticking; if it does, the strike bolt isn't lining up with strike in the jamb.
- If the bolt and strike aren't lining up vertically, check that the hinge screws are tight.
- If only the key side is sticking, try another key to confirm one key isn't excessively worn down.

Powdered Graphite

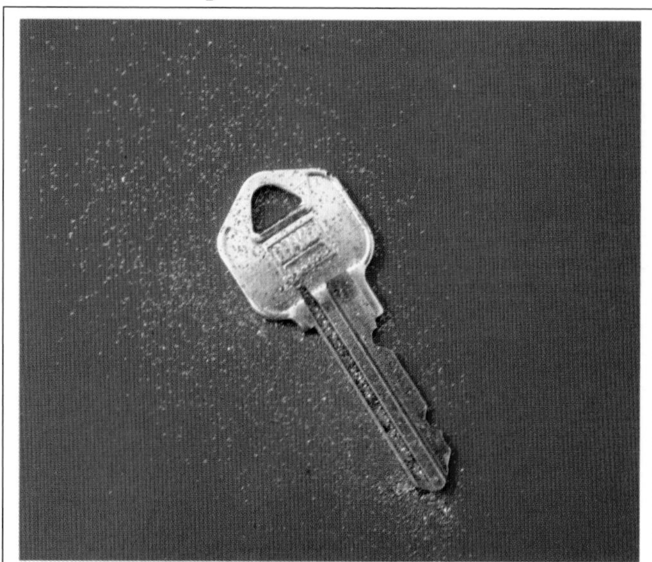

- If the lock sticks with the door open, spray the inside with powdered graphite, available at hardware and auto parts stores.
- Squeeze the powder into the lock through the key opening and at the latch and strike bolt a few times to distribute the graphite.
- Spray additional graphite on the key, and work it in the lock as well.
- If powdered graphite isn't available and the lock must be lubricated immediately, spray silicone in the key opening and catch.

A little detective work, patience, and the right lubricant will solve most of your lock problems. This is an easy project for a homeowner and doesn't require any additional professional help.

Sometimes the plastic liner inside the strike causes the door to stick.

YELLOW LIGHT

Don't mix lubricants (silicone and graphite). You'll get a sticky mess and make things worse. Out of graphite? Rub the key with some pencil graphite (the "lead" in a lead pencil) and work it into the lock or grind the graphite into a powder and liberally coat the key with it.

Door Strike

- The door strike fits into the door jamb, and the latch and the slide bolt fit into the strike to secure the door.
- Check inside the strike for a plastic liner, which can sometimes cause sticking, and remove it.
- You might need to file down the strike until it matches up with the side bolt.
- If filing isn't practical, move the strike slightly. Enlarge the hole in the jamb with a wood chisel and drill the screw holes again.

Disassembling

- Many standard locks come apart by removing screws that pass through the lock assembly and are found near the interior door knob.
- Older mortise locks have screws near the interior door knob and on the door edge where the strike bolt moves in and out.
- Remove any fasteners securing the lock and pull the lockset from the door.
- Look for broken parts, clean out any dirt, and lubricate with graphite before reassembling.

HOLES IN PLASTER

Fixing plaster holes is not as difficult as it looks and doesn't need professional help

Plastered walls are typically made up of three coats of different types of plaster, making them very sturdy. As hard as plaster is, swinging door knobs, removing shelf brackets, and things that go bump in the night can all leave holes to be repaired. Fortunately, we don't have to repair with three coats of plaster any longer. Plaster mixes and fillers are readily available at home centers for one-part repairs that any novice can do.

Plaster dries fast, so give yourself five minutes of tooling time per batch and mix only enough that can be worked in that amount of time. Plaster is too hard to easily sand and is meant to be mixed and smoothed quickly.

Removing Loose Plaster

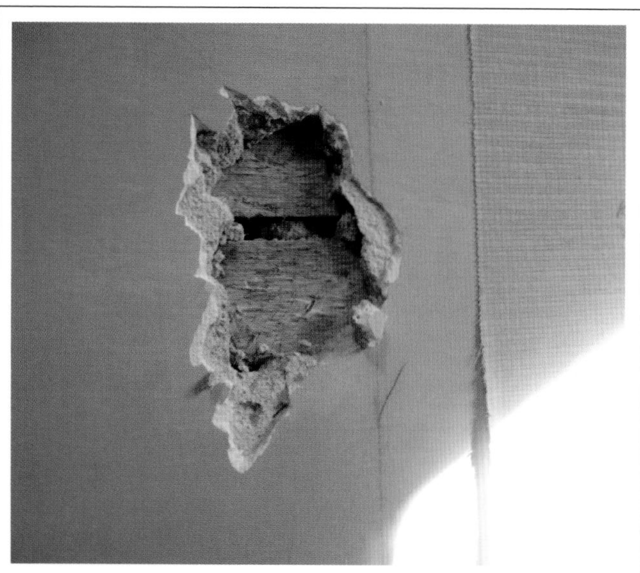

- Place a drop cloth on the floor, and clean out all loose plaster with a putty knife or old screwdriver.
- Dig to the lath using an old screwdriver or the point of a putty knife.
- Break plaster away until solid plaster is found without crumbling—be cautious, it's easy to remove too much.

Clear Away Dust

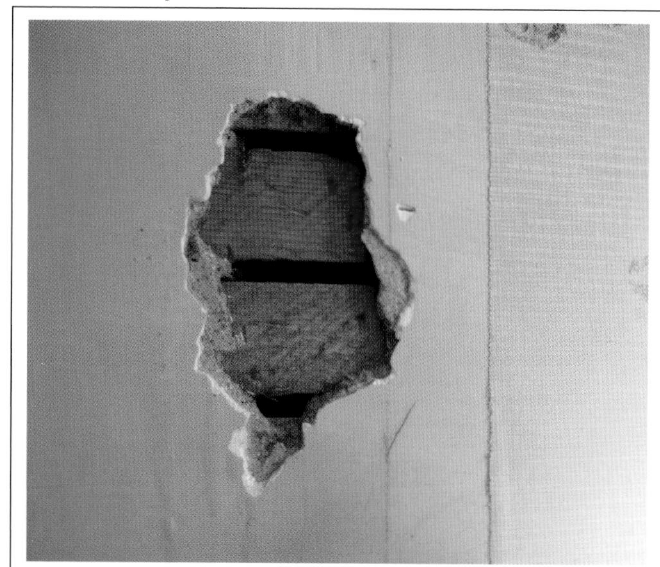

- Brush the dust and loose bits off with an old paintbrush or vacuum off.
- Even small holes can have more loose plaster around them than you might suspect so clean these thoroughly.
- Spray water on the exposed plaster, lath, and surrounding edge of the old plaster.

Small repairs done with plaster mixes are indistinguishable in appearance and hardness from the surrounding plaster. And for larger areas, depending on the size, traditional plaster might be called for. An option for a hole larger than 2 inches across is to cut a section of drywall and fasten it to where the plaster is missing or removed from the original wall. Then skim over the area with drywall compound as a final finish.

MAKE IT EASY

Mixing a small amount of vinegar—about 10 percent of the required liquid—with dry plaster mix in lieu of water will slow down the drying and allow more time to smooth it out. Before mixing, be sure the area to be repaired is completely ready, with all loose plaster removed.

Smoothing It Out

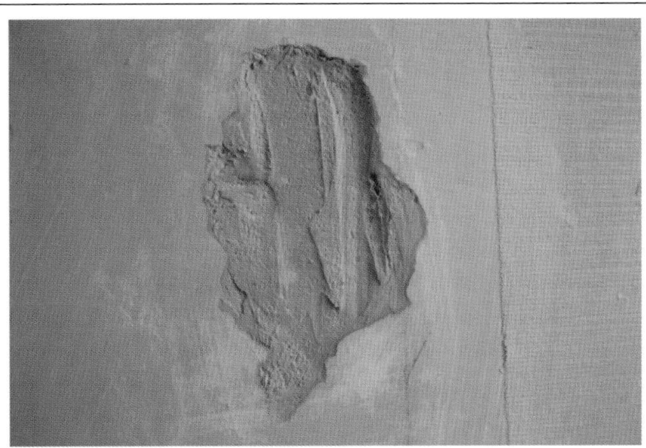

- As many as two or three layers of plaster and other fillers might be layered, depending on the size of the hole being repaired.
- Allow each layer to completely dry—most are fast drying, twenty minutes or less, depending on how thick a layer is applied.
- An experienced plasterer can obtain a smooth finish using only plaster without sanding.

If you're not a pro, apply a final coat of joint compound or Spackle as both are soft materials that can be sanded.

- Apply a small amount of joint compound several inches beyond the patched area.
- Joint compound takes longer to dry than plaster fillers.
- Wipe the dust away from the patch.

Priming and Painting

- When dry, either dry sand with 100-grit paper or wet sand with an old sponge.
- Prime and paint the area to match the surrounding wall—almost any interior primer will do, including fast-dry types, as these patches are normally small and can be quickly coated.
- Wipe any dust from the patch.
- Prime with a foam brush, which does not leave brush marks.
- When it is dry, touch up with one or two coats of paint.

REPAIRING PLASTER CRACKS

A bit of movement, and plaster can crack—here's how to fix it

Plaster is hard, but it can still crack. When the wood framing under it moves or shifts a bit from earth movements or expansion and contraction from unexpected moisture, the plaster moves with it. Whereas wood is flexible and has room to move, plaster does not, so cracks develop.

Plaster cracks often look random, but they have logic of their own. It all depends on what moved underneath which section of a wall or ceiling. And these cracks can be long. Simply running some Spackle over them, a far too common approach, doesn't fix them, nor does the smear of Spackle look at all attractive.

A plaster crack should be opened up and made larger, preferably down to the underlying lath, and then filled. A properly repaired crack will last. You might get new cracks over

Widening the Cracks

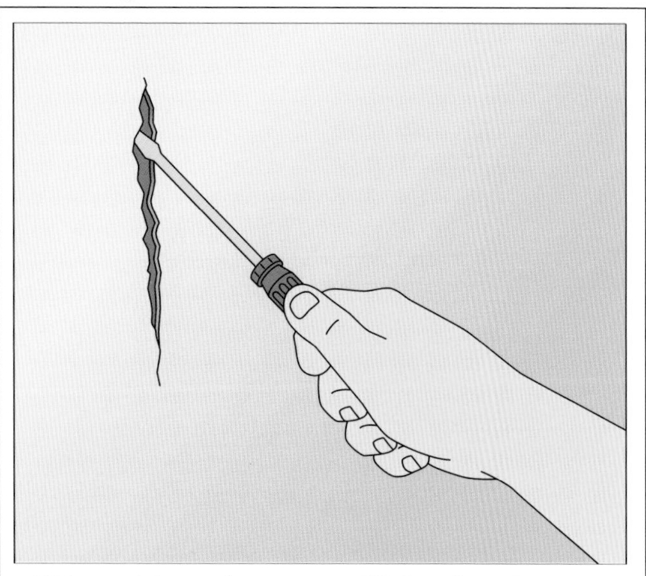

- Widen and dig out the crack using an old screwdriver or the point of a stiff putty knife.
- Some prefer to apply fiberglass drywall tape over the crack and "mud" it over with several layers of drywall compound.
- Filled cracks might return, but taped cracks require more finesse applying the drywall compound.
- If you don't mind lots of dust, use a drill with a grinding wheel attachment to grind all the cracks to the lath.

Applying Fix-All

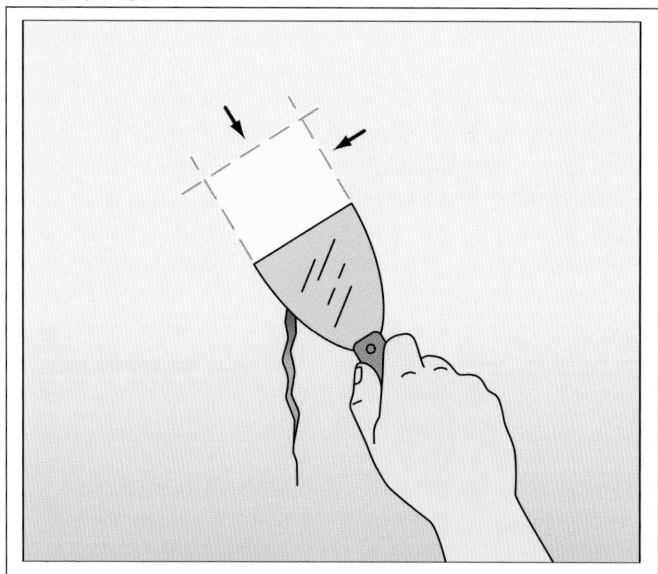

- After the crack has been cleaned out, wet it, pack it with plaster filler, and smooth it with a putty knife.
- Narrow cracks can usually be filled with one application of filler, while wider cracks require two applications.
- Taped cracks call for two to three thin coats of drywall compound, spread thinly about 6 inches beyond each edge of the tape.
- Don't rush sanding the joint compound—sand it only when it's completely hard.

time, but the repaired cracks should mostly stay closed.

Some cracks are so fine you have to decide if they're worth repairing or simply painting over. If you paint them over, there is some risk they will enlarge later with future wood expansions, so weigh your options—repair now or maybe repair later.

ZOOM

Plaster is forced into metal or wood lath, which anchors it as it dries. When plaster cracks, it can pull away from the lath. Special plaster washers are available that, when screwed against the plaster and into the lath, help secure loose plaster without removing and replacing it.

Applying the Final Coat

- Determine if you need to finish coat the crack by its appearance.
- When dry, if the filled crack needs another coat of filler, apply a thin coat of drywall compound, and sand when dry with 100-grit paper.
- After the last drywall compound layer is sanded over the taped repair, run your hand over the patch for smoothness—you don't want a lumpy repair to stick out.
- Be sure the drywall compound completely covers any fiberglass tape; otherwise, it will show through the paint.

Sanding and Painting

- Repaired cracks get primed and painted the same way as repaired holes.
- If a room has multiple cracks, figure on repainting it entirely—otherwise, the painted cracks will stand out too much.
- Prime the crack with a foam brush to avoid brush marks and then apply thin coats of paint until the repair is covered.
- Spread the last coat of paint out thinly away from the repair to better blend in.

DRYWALL HOLES

Fixing drywall imperfections can be done without the help of a professional

Drywall has been around since World War I, but it took the labor and materials shortages of World War II and the demand for housing after the war to kick in the use of drywall over plaster. Unlike plaster, drywall was produced in ready-to-install sheets (4'X8' and larger) of gypsum sandwiched between sheets of durable paper and was nailed directly to framing lumber. Installing drywall was and is a lower skilled trade than plastering, although it still calls for strength and speed.

The joints between pieces of drywall are finished with several coats of joint compound (or "mud") and paper (or fiberglass) tape. After drying, each coat is sanded smooth before the next coat is applied. The joint compound comes in three

Dents and Holes

- Repair minor drywall dents with Spackle or plaster filler.
- Use a repair kit or fiberglass drywall tape (single or double layer) to cover larger holes.
- Apply three layers of drywall compound, spreading about 6 inches beyond the tape, and sand each layer smooth after it's dried.
- If the wall is textured, apply spray texture after the last coat of drywall compound or hand texture the last coat while still wet with a sponge, roller, or other appropriate tool to match the surrounding area.

Repair Kits

- Drywall doesn't have any lath behind it—there's nothing to hold any filler.
- Self-adhesive mesh and metal screen drywall repair kits are available for holes up to 6"X6" or so.
- Attach the mesh/screen, spread some drywall compound, and allow for drying, sanding, and painting.
- Smaller holes can sometimes be lightly cleaned out and repaired with plaster filler, but holes larger than 6"X6" will need backing installed and a drywall patch.

types: all-purpose, topping, and quick set. All-purpose is available at just about every paint, hardware, and home center store, making it convenient for you to handle your own repairs. It comes premixed in plastic buckets and will do just what its name suggests. It will work for any drywall taping or repair purpose. It shrinks a bit more than the other two and, when dry, takes longer to sand to a smooth finish, but it's ready to use and stores easily.

MAKE IT EASY

Drywall is sold in sheets, normally two at a time, far more than you'll ever need for small repairs. It's worth grabbing some scrap pieces from new home construction sites and storing them for future use. Scrap drywall is often piled and stacked outside these sites and is considered discarded.

Cutting Out the Damaged Area

- With a pencil and straight edge, draw an even square or rectangle an inch bigger in each direction around the damaged area.
- Cut this area out with a utility knife or keyhole saw.
- Place thin strips of wood—plywood scraps will do—inside the wall and behind the drywall and secure them by driving drywall screws through the drywall and into the strips.
- The drywall can also be carefully cut out with an electric jigsaw.

Installing the Patch

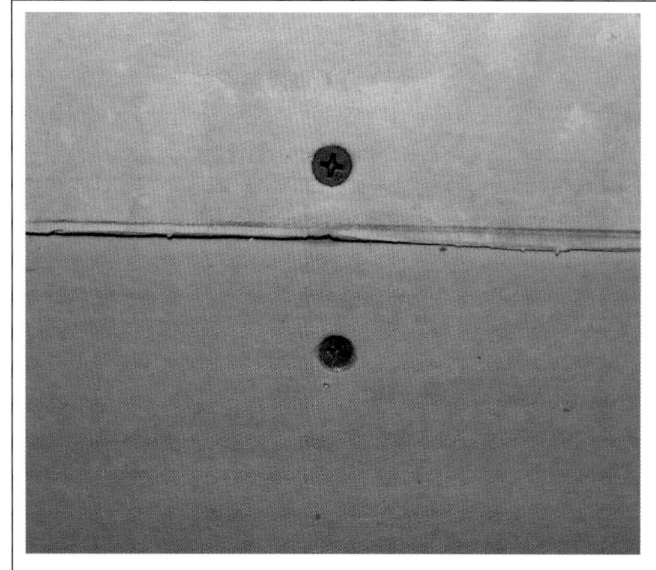

- With the backing in place, a piece of scrap drywall can be installed, taped, and finished.
- Cut a section of drywall the same thickness as the existing drywall to size; glue and screw this patch to the backer supports.
- Tape the edges and apply three coats of drywall compound, extending out about 6 inches, sanding each smooth, and applying a textured finish if needed.
- Prime and paint the repaired section.

DRYWALL CRACKS

Because drywall often cracks at the taped joints, the repairs necessary are different than for plaster

Drywall is a stiff material, but it does move along with the underlying framing it's attached to on walls and ceilings, especially as the framing loses some of its water content and dries out. As the wood moves, it pulls away from the nails or screws securing the drywall. Cracks show through in various ways. Sometimes all you'll notice is a nail or screw head coming through the surface. Other times, the drywall moves at the joints, and the tape and joint compound are torn and damaged.

A drywall repair requires you remove all the loose tape and joint compound from the damaged area. At that point, you can retape and "mud" the crack, but it's faster and simpler to

Cutting Away Torn Tape

- Drywall cracks occur at taped seams where two sheets of drywall come together and the seam is covered with drywall tape and joint compound.
- Cut away any loose tape and joint compound.
- Fill in the damaged area with plaster filler, forcing it into the joint and smoothing it with a putty knife.
- When the filler is dry, recoat with joint compound, texturing if necessary, followed by priming and painting to match the surrounding area.

Tape Repair

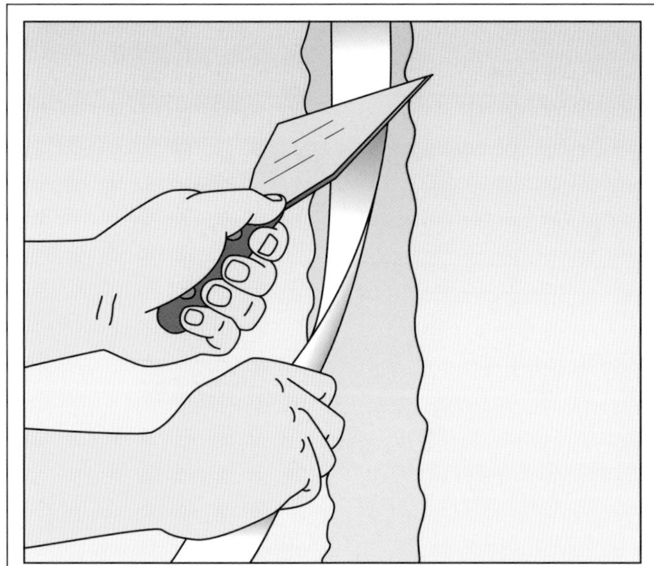

- Use paper or self-adhesive fiberglass tape to retape cracks and duplicate the original construction.
- Remove loose tape and joint compound, apply a thin coat of joint compound, and cut a piece of new paper tape the length of the crack.
- Press the paper tape into the compound with a putty knife and spread a thin layer of compound over the top of the tape until it's smooth, extending out about 6 inches from the tape edges.
- Fiberglass tape is self-adhesive and applies directly to the wall board; do not apply it to joint compound.

use patching plaster or other filler. Taping and "mudding" a crack just isn't necessary unless a very long crack is being repaired with a lot of tape damage. Patching plaster will fill and replace any smaller damaged taped areas without compromising the joint or appearance.

YELLOW LIGHT

If your drywall crack is wet, you have a water problem that must be repaired before the crack is repaired. This will probably call for removing a larger area of drywall. Drywall cracked across its face and not at a seam suggests major settling, and the cause should be investigated.

Applying More Coats/Texture

- It's easy to apply joint compound too thickly and end up with a lumpy repair—good taping and "mudding" is an art form that takes practice.
- All coats should be spread thin with a wide finish knife or putty knife.
- Always use a clean knife for spreading—all bits of dried joint compound should be wiped from the edge.
- Sand off any ridges, high spots, and rough edges with 100-grit sandpaper (the last coat should require little sanding).

Nails

- If a drywall nail pops out, drive it back in with a hammer and nail set.
- If a screw shows, lightly tap a Phillips screwdriver into the head and tighten it by hand.
- Screw the drywall about 3 inches above and below the loose screw/nail.
- Press a small amount of Spackle or drywall filler over all nail/screw heads with a putty knife, sand the filler when it dries, and prime and paint with an artist's brush for the least visible touch-up.

CRACKED CONCRETE FLOORS

Fill serious concrete cracks to keep the area safe and prevent water from seeping in

While it's one of the hardest materials in housing construction, concrete is still susceptible to cracks. This isn't necessarily a sign of weakness, but it's in the nature of the material and its installation. Concrete shrinks some as it cures and also swells and shrinks as it absorbs and loses moisture, although these changes won't always cause cracks.

Water not only turns dry concrete mix into pliable, usable material but also chemically reacts with the cement in the concrete mix causing the hardening process. Without water, concrete wouldn't cure and harden properly. Curing can go on for years, which explains why cracks can show up any time. How the concrete was mixed and poured, the weather con-

Caulking Hairline Cracks

- Small hairline cracks and slightly larger ones can be filled with latex concrete caulk with silicone. Caulking cracks brings no guarantees, but when it works, it's a simple repair with good results.
- Vacuum the crack out before caulking, following the directions on the caulk tube.
- Press the caulk into the crack and scrape it off the concrete surface with a putty knife, leaving only a thin film.
- Reapply if needed to fill the crack to its top.

Cleaning Out Larger Cracks

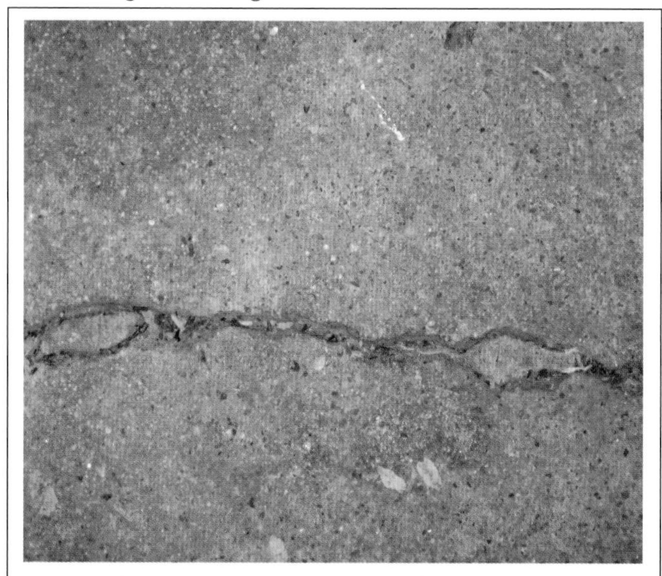

- Wear safety glasses to remove loose material with either a hammer and cold chisel or an old unwanted screwdriver, making the bottom of the crack wider than the top.
- If dust isn't a problem, use a grinding wheel attached to a drill as a fast way of clearing out a crack.
- Vacuum up all loose debris and dust.
- Apply a suitable concrete repair mix and press it down completely tightly into the crack.

ditions, the amount of water in the mix, and ground preparation all affect the degree of cracking in a poured concrete slab. Even under the best of conditions, shifting soil under concrete creates conditions that cause cracks.

Cracks can be serious or merely cosmetic and nonstructural. Hairline cracks are good examples of cosmetic, leave-them-alone cracks. More noticeable cracks, however, should be filled to improve appearances, to ensure safety (if the edge of the crack is a tripping hazard), and to keep water from seeping in and worsening the crack.

RED LIGHT

Cracks in a concrete floor are one thing, but deep cracks in concrete walls can be structural issues and should be reviewed by an engineer or contractor. Your home might not be endangered, but you don't want severe cracks to progress if simple repairs are possible.

Cleaning Out a Hole

- Clean holes in concrete of all loose, damaged material using a hammer and old screwdriver or cold chisel.
- Clean all exposed sides of the hole with a wire brush, vacuum out the loose debris, wet the area, and fill it with latex cement, a two-part product (follow the instructions on the can), pressing it tightly.
- Slowly build up the material as you fill the hole.
- Smooth the final layer with a broad putty knife.

Large Holes

- Depending on the hole location and whether cars drive over the repaired areas, large holes call for different types of concrete mix.
- A concrete mix with gravel added rather than an all-sand mix will withstand the weight of an automobile for driveway and garage floor repairs—ask your supplier for the appropriate mix.
- A large hole can require more aggressive clean-out, including the use of a sledge hammer, so you might need an expert here.
- Mix the concrete thoroughly according to the package instructions, press it into the hole, and follow the curing instructions.

CRACKED GLASS

You can't repair a glass crack, but you can replace the glass itself

Glass is a mix of sand, lime, and soda, all heated together until liquefied and then poured to form different shapes. It's the best material for windows and should remain crack-free for a lifetime if we don't traumatize it by throwing baseballs through it, slamming windows closed, or running a high-temperature heat gun against it. But we do traumatize glass and occasionally need to replace it while avoiding future traumas.

Replacing older, single-pane glass is relatively simple. The glass is available at most hardware stores or local glazing shops at a reasonable cost, and removing the glass doesn't require any special tools. It's also somewhat intuitive: You can see by looking at the glass how it's secured to the sash.

Aluminum Windows

- Broken single-pane glass in steel or aluminum windows can be held in by glazing compound (putty), a sealant, metal clips, or by a metal frame, screwed at the corners.
- Spring tension clips, which insert into the frame, help hold puttied glass in place.
- Remove the broken glass; when measuring for replacement piece, buy a piece about 1/8 inch shorter in each direction from your measurements for easy fitting.
- Replace the damaged sealing product (e.g., putty) with the same material when installing new glass.

Wood Windows

- In older wood-sash, single-pane windows, glass is held in with glazing points—small sharp fasteners pressed against the glass—and glazing compound, which seals the glass to the weather.
- Chipping the old glazing compound out is the toughest and most time-consuming part of the job—all of the compound must be cleaned out.
- Measure the new glass and remove 1/8 inch from each measure before ordering new glass.

Repairing a modern insulated glass unit isn't so intuitive, so sometimes the aid of a professional glass contractor might be needed. That said, modern glass is less likely to have cracks than old glass in part because of its age (it's had less exposure to traumatizing incidents), improved manufacturing methods, and thickness. Any cracked glass can be repaired, but unless the glass is in danger of falling out, repairs aren't critical.

YELLOW LIGHT

Unsure if your crack might get worse and possibly fall out, but you can't repair it right away? Tape both sides of the crack with clear mailing tape, which stands out less than other tapes. The glass is less likely to fall out and injure anyone walking below it.

Sealants

- Glazing compound was the standard sealant for single-pane glass for years and is still used today for repairs.
- Remove, knead, and soften the entire can of glazing compound before using.
- Glass is sometimes secured to metal sashes with glazing tape and then sealed with silicone caulking—it all depends on how the original unit is glazed.
- For paintable sealants such as glazing compound, follow the manufacturer's instructions for curing times before painting.

Fixing Insulated and Single-Pane Glass

- Single-pane glass can be cut in any glass shop or hardware store that sells glass, but insulated glass has to be made to order unless your window manufacturer keeps your desired size in stock.
- Each window manufacturer has different assembly methods and fasteners.
- If there is no apparent way to remove a sash with broken glass, contact the manufacturer for directions or call a glass installer (glazier).
- In some cases, the entire sash might need replacement if the glass cannot be removed.

HOW YOUR SYSTEM WORKS

Learn the ins and outs of your system to prepare for maintenance and safety in the future

We don't see electricity, but we see the results: It runs appliances, turns on the lights, and displays television shows. We pay attention to it only when it isn't working or when the bill comes. Modern electrical systems are very reliable and safe. The only changes that occur are updates to the electrical code and a never-ending supply of new devices to plug in and turn on. Older systems are often safe as well, but problems occur when changes are made that don't meet code and safety requirements.

The United States Fire Administration claims that home electrical problems account for almost 68,000 fires a year, most from faulty wiring and misuse of cords and plugs—

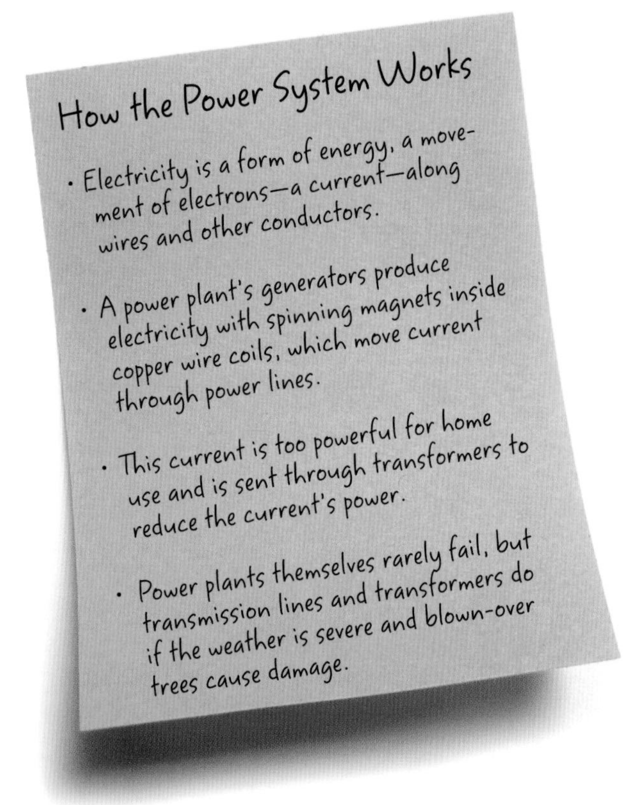

The Service Panel

- A service panel, or fuse box, distributes electricity to loads—lights, appliances, outlets, etc.—and protects wiring. The more circuits, the better since too many loads or loads demanding a lot of electricity on a single circuit can cause problems.
- Fuses and circuit breakers stop electricity in overloaded circuits.
- New full-size homes typically have 200-amp service panels; fuse boxes are no longer installed in homes.
- In old homes, upgrading to a service panel is a good idea.

especially during December when indoor lighting use increases. Understanding the limits of your system and basic maintenance will keep you safe and make the most efficient use of your electricity.

Electricity Terminology

- Electricity has its own terminology:
- The electricity you use is measured in watts—a 100-watt bulb consumes 100 watts of electricity.
- A kilowatt is 1,000 watts.
- You are billed in kilowatt hours—the amount of electricity used by ten 100-watt light bulbs in one hour, for instance, is a kilowatt-hour.
- An amp is an amount of electricity.
- Voltage is electrical pressure.
- Circuits are measured in amps.

GREEN LIGHT

If you're uncertain about an older electrical system, hire an electrician to do a safety check. It typically takes about an hour or so for the inspection. Consider any recommendations for suggested safety upgrades, but foremost be sure it's safe enough as is and understand any limitations on usage.

Knob and Tube Wiring

- Knobs and tubes are ceramic hardware once used when installing and attaching wiring. Knob and tube wiring typically has rubber and black cloth insulation around single wires.
- There's no ground wire (a safety feature in new systems) in knob and tube systems.
- Modern wiring contains all the required wires in one cable, each wire having its own plastic insulation around it.
- Romex is both the trade name and common name for modern insulated wiring.

FUSES & CIRCUIT BREAKERS

Protect yourself by getting familiar with your fuses, wiring, and circuit breakers

The wires in your house are sized to carry a certain amount of electrical current. Too much current, and wire can overheat and cause a fire. Fuses and circuit breakers limit how much current can travel through wires by shutting down the circuit. They are the main safeguards of your system. Without them, you could plug in a dozen small appliances in your kitchen on a circuit designed to run just one or two, and the current would keep flowing until the wire becomes red hot.

When you demand too much current by trying to run a toaster, blender, and coffeemaker at the same time on one circuit, the breaker trips, a fuse "blows" or "burns out," and the flow of the current stops. They act like on/off switches that

Fuse Box and Fuses

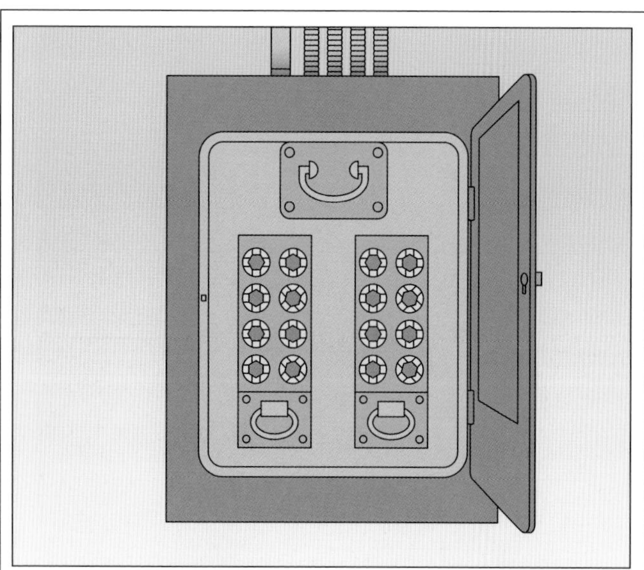

- Most household fuses are glass screw-ins with thin metal strips in them; if too much current passes through them, the strips melt and the fuses "burn" out.
- Fuses are rated by the amperage that passes through them—15 amps, 20 amps, etc.—which determines the size of the wire they can protect.
- A fuse will last indefinitely until it's overloaded, which is more than its circuit can safely handle.
- Always keep extra fuses on hand for quick repairs.

Service Panel and Circuit Breakers

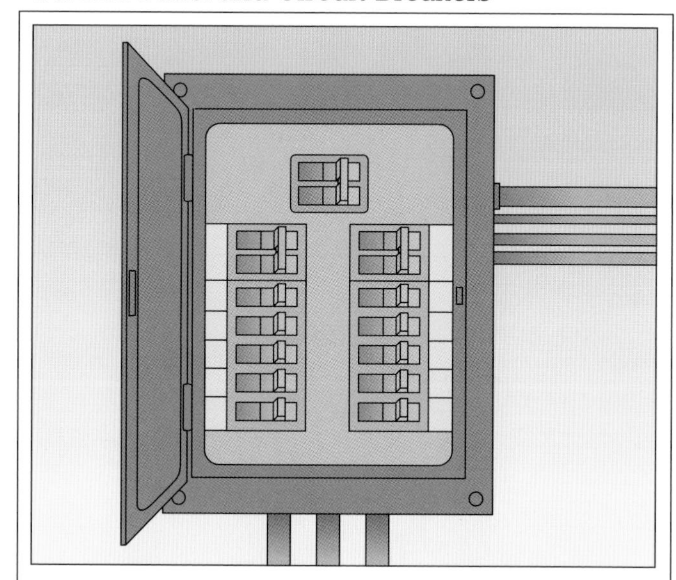

- A modern service panel has circuit breakers that "trip" to stop the flow of electricity in an overloaded circuit.
- A circuit breaker can conveniently be reset and used indefinitely, unlike a fuse, which needs replacement after it burns out.
- Circuit breakers are also rated to protect certain sizes of wires—a 15-amp breaker protects 14-gauge wire, a 20-amp protects 12-gauge wire, etc.
- Circuit breakers are found in grounded electrical systems.

have to be reset (circuit breaker) or replaced (fuse). Never, ever change out a fuse or circuit breaker with one of larger amperage because the power keeps shutting down on a particular circuit. The wires, circuit breakers, and fuses are designed to work with each other. Tamper with this, and a fire can result, which often isn't covered by insurance. So be sure to understand and work correctly within your particular circuit system.

RED LIGHT

Old fuse systems can be perfectly safe, provided each circuit has the correct fuse in it. A past resident might have replaced a low-amp fuse with a higher amp fuse, and this is a fire hazard. Except for large appliance loads, circuits for lights, bedrooms, and common rooms will normally be 15 amps.

Replacing a Fuse

- When replacing a blown fuse, remember these precautions:
- Unplug and turn off any loads on the affected circuit.
- Remove any jewelry from your hands and wrist.
- Use a flashlight and plastic fuse tongs, which do not conduct electricity; be sure the floor under the fuse box is dry.
- Keep one hand behind your back and the other on the fuse—a second hand can come in contact with the fuse box and form a complete path for an errant electrical current.
- Replace the burned-out fuse with one of the same amperage.
- The last item plugged in before the fuse blew should not be plugged in again, even if the cord is good—this device drew too much power and caused the fuse to fail.

Resetting a Circuit Breaker

- Unplug and shut off all loads on the circuit. Be sure the floor under the panel box is dry.
- At the panel, find which breaker has tripped—it won't be lined up with the other breakers.
- Push the breaker into a complete "OFF" position and then push to "ON." It will click when it's on.
- Check all plug-in loads for frayed cords.

OVERLOADING A CIRCUIT

If you're not mindful of your system, you can overload it and cause a problem

Modern home electrical systems normally dispense enough current in convenient locations that overloading doesn't happen very often. In addition, circuits are designed with a safety factor to accommodate sudden surges (demands for power from energy-gobbling appliances such as vacuum cleaners and toasters). At the same time, homeowners have so many toys and electronics demanding power that one too many can pull just enough current to throw the circuit over its limit. You will need to rethink the loads on a given circuit to better distribute them to other circuits.

The first step is figuring out what each circuit runs—which lights, which receptacles, any appliances. Be wary of extra

How Much Load on a Circuit?

- A circuit is a section of your electrical system and can handle only a certain amount of current before it shuts down.
- A 15-amp circuit is designed for 1,440 watts of load and a 20-amp circuit for 1,920 watts of load.
- Every appliance, light, computer, and so on has a wattage rating—add them up for the total watts.
- Kitchens and bathroom are normally wired to carry heavier 20-amp loads. Hallways, dining, living, and bedrooms usually have 15-amp circuits for lighter loads.

Overloading

- Plug in a toaster while the microwave, blender, and food processor are running on one circuit, and you'll then find nothing is running—the circuit is overloaded.
- Shutting down is a safety measure to protect the wiring, nothing more.
- If a circuit continually works until you add a new load, remove that load or an equal one, and recalculate your load total.
- Any new load can cause an overload, but typically it's a small appliance, tool, or space heater.

loads feeding into the system, which might lead to overloading the circuits. If all circuit loads aren't turned on at once, it might be easy to overlook an overloaded circuit, so double check every receptacle and light. It must be assumed that all the loads on a circuit will be on at the same time and that collectively they don't exceed the circuit's limit.

Once you understand your system, you can better anticipate future problems and easily avert potentially dangerous situations.

YELLOW LIGHT

It could be that your system is too underpowered for all your loads—100 amps, for instance, when you need 150 amps or more. Other than not running some loads, there is no repair for this other than upgrading the system. If you continue to lose power to certain circuits, talk with an electrician about a bigger system.

Power Surge

- When a motor turns on, it needs more power initially during the surge than it needs when it's running.
- Common household appliances and tools have surge factors: vacuum cleaners, hair dryers, washing machines, refrigerators, electric saws, drills, air conditioners, and furnace fans, to name just a few.
- Before extending a circuit you must confirm the circuit has enough power to run that load if all other loads are also running.

Making a Circuit Map

- The inside door of your fuse box or circuit panel should have a general list of which fuse/circuit breaker controls an individual circuit—a more specific list helps determine the circuit's true load level.
- Turn on every light in the house, and plug a radio or light into every receptacle to test your circuit.
- Turn off one circuit breaker at a time (or unscrew one fuse) and record what loads shut off.
- Listing every load informs you if your circuit can accept additional loads or whether it's maxed out.

SAVING ELECTRICITY

Keep electricity usage low—and your electricity bills even lower—with a few simple tips

Energy conservation comes and goes, depending on the latest oil or electrical grid crisis and price hike, but with some minor adjustments and a bit more awareness of our electricity use, every household can cut back, save money, and give a stressed infrastructure a break. Get into a daily, weekly, or monthly routine to help drop the cost and usage of electricity.

There are obvious things you can do every day to conserve electricity—turn off lights when they're not needed, run the laundry appliances and dishwasher when full, cook several meals at once in a heated oven—but there are also less obvious habits. Doing an energy audit of your home and lifestyle will surprise, inform, and guide you to new considerations

Computer and Power Strip

- According to the Lawrence Berkeley National Laboratory, the electronics in the average home consume over 50 watts an hour even when shut off, but this usage can be minimized.
- Shutting a PC off at the power strip will stop all electronic activity; newer models can handle many on/off cycles without damaging the hard drive.
- Use "sleep" or "hibernate" modes for added energy savings.
- Use power strips to shut off other electronic devices to save power as well.

Compact Fluorescent

- A compact fluorescent bulb (CFL) replaces a standard incandescent light bulb.
- ENERGY STAR qualified bulbs use about 75 percent less energy than standard incandescent bulbs and last up to ten times longer.
- A CFL can cost five or six times the price of an incandescent bulb and can save more than $30 in electricity costs over the bulb's lifetime.
- Dimmer switches require a special CFL.

regarding your energy habits and consumption. All those computers and entertainment devices still require electricity even after you've shut them off. And dropping the heating temperature one degree a month to acclimate to a slightly cooler house can bring considerable savings. Every little bit adds up.

MAKE IT EASY

Keeping lighting usage down is great, but you can't live in the dark. Install timers and attach them to low-wattage lights for both security and safety (in case you have to get up in the middle of the night). Automated systems for multiple devices programmed with a PC are available as well.

ENERGY STAR

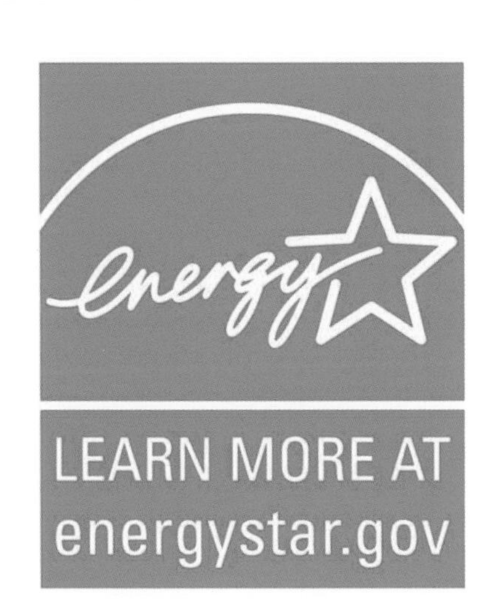

- ENERGY STAR is a joint voluntary labeling program of the U.S. Environmental Protection Agency and the U.S. Department of Energy.
- The ENERGY STAR label indicates lower energy usage and more efficient operation than comparable models.
- Products that pass the program's energy standards perform as well as or better than comparable models while being more energy efficient.
- Via its website and printed materials, this program offers energy and building assessment tools as well.

Hard-Wired Energy Savings

- When rewiring a home or a single room, there are opportunities for hard-wired energy savings.
- Instead of depending on a power strip, install switch-controlled receptacles that you can conveniently shut off using a wall switch.
- Install low-wattage wall and ceiling lights on a separate switch as options for lower energy ambient lighting.
- Use built-in LED night lights to provide just enough illumination through low watts for late-night walking around.

LAMP, CORD, & PLUG REPAIR

Before throwing out your radio, light, or other device, try repairing it

When it comes to repairing lamps, cords, and plugs, take a lesson in frugality from our grandparents and great-grandparents: Try a repair first before you toss it in the garbage. A common repair with small electrical devices is replacing a maimed cord or plug. Hardware stores always stock replacement plugs for lamps and radios and other small devices as well as for heavy-duty extension cords. Replacing a plug is a very straightforward homeowner repair.

Replacing a lamp or power tool cord can be a bit trickier—some disassembly is involved, and you don't have a lot of extra room to work—but given the replacement cost of an

Rewiring a Lamp

- New lamp cords, plugs, and sockets are available at hardware and lamp stores.
- For less than $10 in materials, you can restore a lamp to full use.
- Replace all the parts in an old lamp; don't bother keeping the sockets unless they're in perfect condition.
- If the bulb isn't lighting, unplug it and pull up on the brass tab at the bottom of the socket—it needs to make good contact with the bulb for it to work.

Repairing an Extension Cord

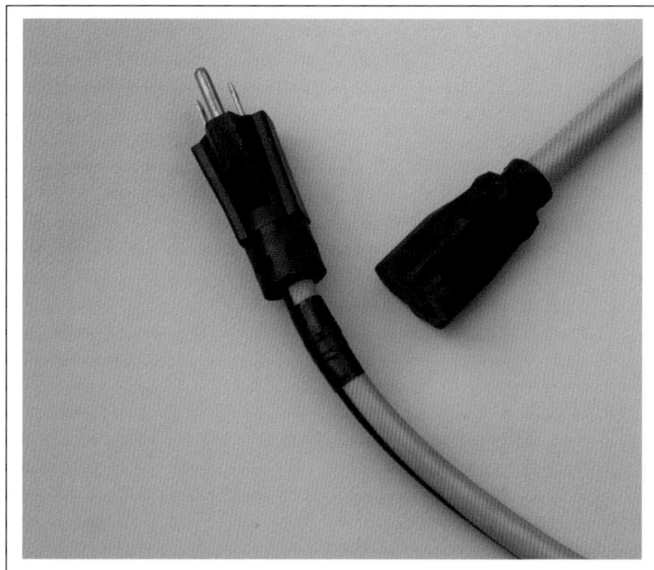

- Extension cords can get damaged by crimped wires, tears, and broken ends.
- Both the male and female ends are replaceable.
- Small tears can be patched with electrical tape but should then be monitored for future wear and tear.
- Heavy-duty extension cords have three wires that attach to the cord ends: green wire to the green screw, white wire to the silver screw, and the black wire to the darker or brass screw.

antique lamp or a good power tool, it's well worth doing. Feeling a little out of your league? Neighborhood hardware stores sometimes do cord replacements.

Torn extension cords can be given a new life by cutting them at the damaged area and installing a plug on the closest end. The cord will be shorter, but there's always a use for short extension cords. This repair is safer than taping over torn insulation (which should be avoided).

YELLOW LIGHT

Be sure the plug you purchase is rated for the cord it will be connected to. Do not use a lamp plug designed for 16-gauge wiring connected to a 12-gauge extension cord. All components in an electrical system must match up in terms of amperage (the amount of current they can safely carry).

Replacing a Plug

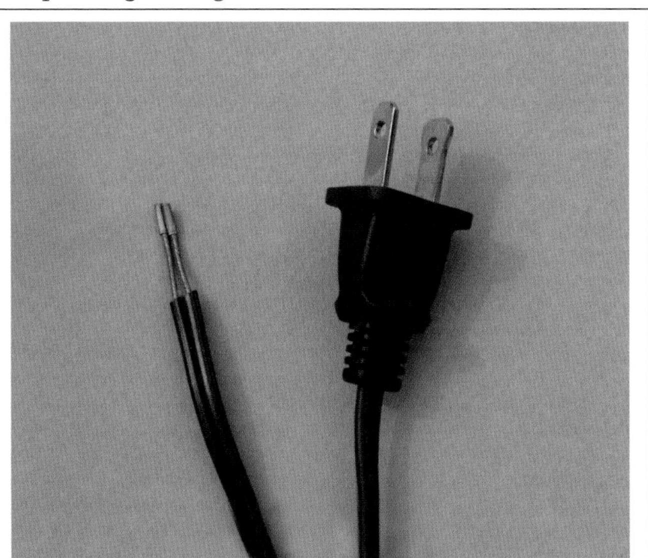

- Always replace cracked or damaged plugs.
- Double-insulated appliances, tools, and other small electrical equipment normally use two-pronged, ungrounded plugs—do not replace these plugs with three-pronged plugs.
- Snap-on replacement plugs hold the electrical cord in a clamp and do not require screwing the wires tight.
- The neutral wire in the electrical cord is wrapped in ribbed insulation and attaches to the silver screw (the wider of the two prongs); the other wire attaches to the brass screw.

Avoiding Damage

- Pull an electrical cord out only by the plug, never by yanking on the cord.
- Don't exceed the maximum wattage light bulb for lamps.
- Never run an electrical cord under a rug or carpet—the insulation can rub off, and a fire can start under the right conditions.
- An extension cord's wire gauge must match with its power demand—a 16-gauge cord cannot accommodate the same load as a 12-gauge cord.

RENEWABLE ENERGY

Decide if the biggest renewable energy installations in history will work for you

Renewable energy has been around in one form or another for centuries. Water-powered wheels and wind-powered pumps go back thousands of years. Solar energy, the most promising for residential use, was captured by the Greeks and Romans who oriented buildings to capture sunlight for radiant heat in the winter but not so in the hotter summer months. During the energy crisis of the 1970s, alternative energy sources were seriously considered in America but fell out of favor due to unproven, badly installed technology, and a drop in the price of oil.

Today, more renewable energy installations than ever before are showing up all over the world, some of them quite massive.

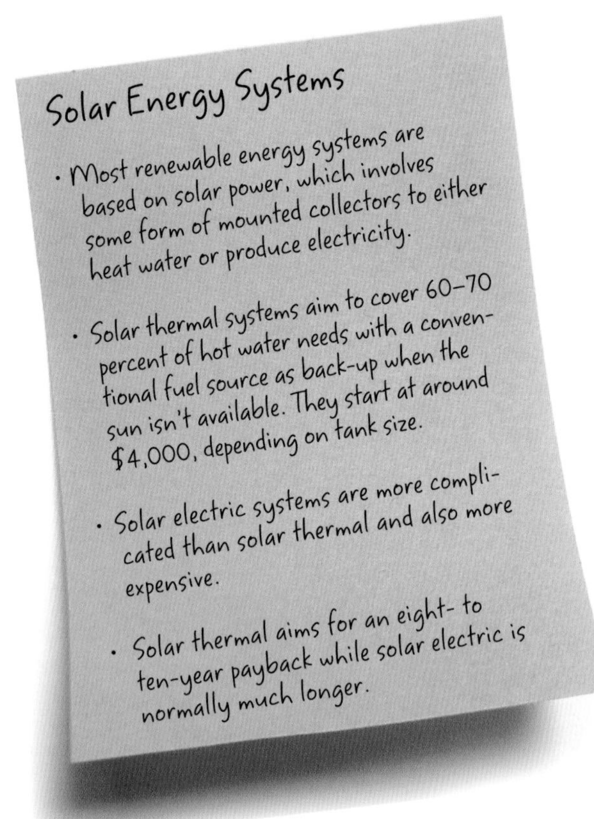

Home Renewable Energy System

- A renewable solar electric system includes a solar array or series of solar collectors/panels, components to convert the solar energy into usable household electricity, and batteries to store power.
- This system calls for more maintenance than a utility-supplied system and may require the help of an expert.
- Battery back-up provides power in the event there is no solar or utility power available due to bad weather.
- Some systems include gas or diesel generators as back-up energy source.

Residential systems are more reliable, better integrated, and even better looking. Nationwide, however, renewable energy accounts for less than 2 percent of all generated electricity.

Is it worth installing a solar energy system in your home? Will you ever recover the cost? What about tax credits? Some systems sell power back to your utility, gaining some small financial advantage. Increasingly, in tax credit-friendly states such as California, new construction offers solar installations.

There are a lot of factors to consider when it comes to renewable energy, but it's definitely worth looking into.

RED LIGHT

Solar electric installations require experienced contractors familiar with the components and a track record of satisfied customers. This is not a job for any electrician or roofer. Look for an installer certified by the North American Board of Certified Energy Practitioners (NABCEP) and ask to talk with previous customers.

Components

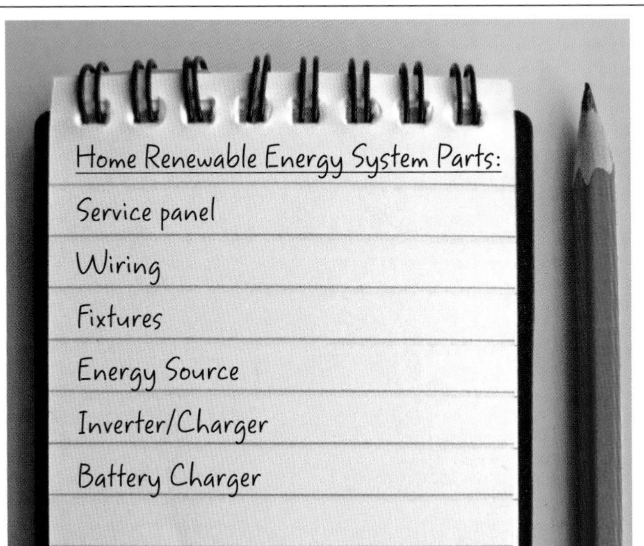

- A home renewable energy system starts with the home's service panel, wiring, and fixtures.
- Additional components include a source of energy—usually a solar array—an inverter/charger for converting the array's DC power to AC, a battery charger for systems with battery backup, and separate panels with their own circuit breakers.
- The system can be adjusted as conditions and power demands change for maximum efficiency.
- It's important that the system is sized properly for the amount of power required.

What to Know Before Installing

- Most climates can support a degree of renewable energy, but some locations are superior in terms of sunlight availability and temperature.
- Rooftop installations must be accurate to avoid leaks and to assure the roof can carry the weight of the array.
- Roofing shingles that double as solar panels are available but do not perform as well as a separate array.
- Renewable energy installations are subject to local and national electrical codes—local inspectors are not always knowledgeable about the applicable rules.

KNOW YOUR MICROORGANISMS

Identifying unwelcome life-forms in your home makes it easier to get rid of them

You're not the only one living in your house. Underneath and on top of the surface are microorganisms, wind-blown seeds, and fungi that can grow out of control. Normally, we all coexist thanks to soap and water, good ventilation, and dry conditions. Sometimes, the relationship breaks down, and you find black stains on a bathroom ceiling or green fuzzy stuff growing on the roof. When it's really bad, you get true rot, the eating away of wet wood because of an unseen water problem. While you can't eliminate the source of spores and fungi, you can control them in your house.

All of these invisible-to-the-eye freeloaders feed off wood, paper, drywall, and some paints, and they can wreak havoc

Mold

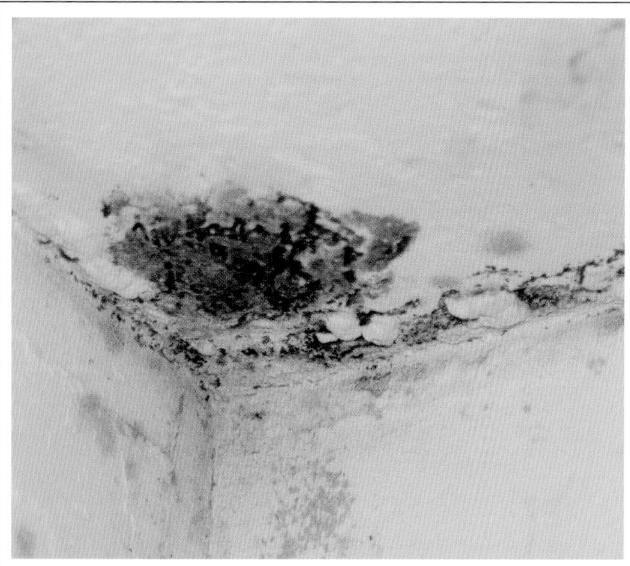

- Mold is a fungus, a type of plant that produces airborne spores that are everywhere and survive only in damp, moist areas while living on organic materials such as wood or the paper covering on drywall.
- You must fix the moisture problem that supports the mold.
- Good housecleaning practices, including keeping outdoor decks swept clean, can help keep the presence of mold to a minimum.
- Mold, which comes in different colors, can irritate allergies and asthma.

Mildew

- Mildew, also a fungus, is a type of mold—the terms are often used interchangeably.
- Growth occurs in areas of high humidity, poor ventilation, and warm temperatures.
- Bathrooms are a favorite for mildew growth, and the only effective cure is good ventilation after cleaning and removing any existing mildew and sealing the surfaces it grew on.
- Synthetic fabrics resist mildew, but cotton, linen, rayon, silk, wool, leather, paper, and wood do not.

on your home. On roofs, moss gets an anchor hold on the edge of shingles and never lets go. Rot can cause structural damage that also needs to be addressed. In extreme cases of interior mold, your health can be endangered, and the mold should be professionally removed.

Understanding these biological menaces will help you eliminate them and protect your home from future problems.

YELLOW LIGHT

Be sure to wear protective latex gloves when scrubbing away mold and mildew. You don't want to chance a skin reaction to the fungi nor expose yourself to the strong soaps and bleach used during the cleaning. If you're especially sensitive, an appropriate respirator might be needed as well.

Moss

- Moss is a plant without a normal system of roots, stems, and leaves that lives in moist locations and is spread by air-borne spores.
- The edges of wet roofing shingles attract moss, and as it grows, the shingles stay wetter longer and deteriorate.
- Moss is very hardy and survives drought conditions as well as conditions in Antarctica.
- When moss is found in lawns, it's an indicator of overly wet conditions, too much shade, acidic soil, or poor watering practices.

Dry Rot

- "Dry rot" describes crumbling, deteriorated wood that had been a food source for fungi when it was wet, but once the wood dried, it became soft and powdery.
- Some wood-rotting fungi require the wood to be wet, while other strains settle on dry wood and obtain water from a source several feet away.
- The fungi break down the wood's cell walls and weaken them.
- The moisture that enabled the rot must be controlled in order to repair its damage.

INTERIOR MOLD & MILDEW

A good cleaning and ventilation are your main weapons against these pesky fungi

The terms *mold* and *mildew* are often used interchangeably for nasty green stuff you never want to see again. In fact, each is a type of fungi, microorganisms spread by spores released from plants. In the outside world, they're a good thing, as they help to break down organic material. Inside, however, they're an unsightly nuisance at best and a health concern at worst to those sensitive to them. Both like wet, warm environments and food—walls, ceilings, fabrics, and carpet—all of which can be wet long enough to support mold and mildew life.

The obvious candidates for mold and mildew are poorly ventilated bathrooms and kitchens as well as air condition-

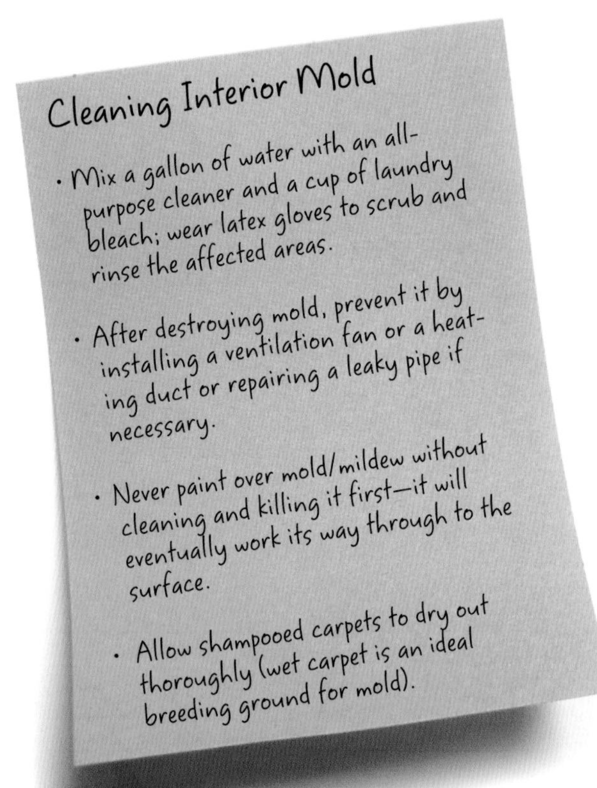

Heat and Ventilation

- Controlling interior temperatures and humidity levels throughout the year goes a long way toward controlling mold and mildew growth.
- Closets on noninsulated exterior walls have a higher humidity level in cold weather—leave their doors open to keep warm air circulating and clothing dry.
- Observe your windows for mold growth in the winter, and increase your furnace temperature to reduce window condensation.
- Be selective about storage in an unheated basement—clothing and books can become mildewed under the right conditions.

ing units and dehumidifiers. Any unseen damp area due to a roof or pipe leak can support fungi growth.

A good cleaning with hot water and a strong cleaner/bleach mix will remove most mold and mildew, but you must correct the conditions that allowed them to grow as well. Cleaning cures the symptoms but doesn't cure the cause. A solution could be a simple matter of turning the heat up higher during the winter months or installing a ventilation fan (or using an existing fan more often and for longer periods of time).

MAKE IT EASY

Keep a squeegee or extra towel in the shower to wipe down tile walls after showering. This helps the grout dry out faster and diminishes the chance for mold to grow. As a plus, your shower is easier to clean when you do your weekly housecleaning, and the sealer on the tile lasts longer.

Ventilation Fan

- A large-capacity bathroom fan will remove moisture fast and is especially effective if set on a timer.
- Any ventilation fan must be ducted to the outside either through the roof or an exterior wall.
- Whole-house ventilation systems are timer-controlled to initiate air exchanges on a user-determined basis.
- Before installing a ventilation fan, confirm whether your existing wiring can be used or whether a new circuit will be required.

Painting a Wall

- Thoroughly clean any surface with mold before repainting to prevent the mold from growing under the new paint.
- Many new paints formulated for exteriors, bathrooms, and kitchens contain mildewcide to resist mold and mildew growth, but always check with your supplier before purchasing.
- Some advise against painting any areas a child could chew or suck on with paint containing mildewcide.
- Inexpensive paints and those containing linseed oil are susceptible to mildew.

EXTERIOR MOLD

Mold can damage the outside of your house, so remove it when you see it

As much as mold likes living inside, it mainly exists outside on the surface of plants. As spores are released, mold finds all kinds of new homes, including decks and siding. Why decks? They're often not maintained especially well, and the typical coating of stain—and its fungicide—isn't long-lasting. As the spores land on a deck, all the exposed wood is more susceptible to the fungi. Decks in damp climates or overshadowed by trees are even more inviting. Regular sweeping and scrubbing with a mild household cleaner go a long way to keeping mold at bay.

Siding can be susceptible to mold growth, too, especially when it's on a shaded side of a house. Shrubs and bushes

Outside Spores Growth Conditions

- The outside air is full of spores and plenty of places to land and multiply in temperatures ranging anywhere from 40 to 100 degrees F.
- Exterior wood that's shaded and/or damp is a likely mold-breeding area—remove or trim any offending landscaping, and the mold should dry up with sufficient sunshine.
- Mold fungi do not structurally attack wood; they do affect paint finishes and general appearance.
- Be sure any exterior paint you apply has added fungicide if mold is a problem.

Landscaping

- Landscaping too close to a house, especially on the north side, will not allow enough heat and light to dry out wet siding and windows.
- Consider leaving at least 18 inches between the plants and the house.
- Cut down any tall bushes or shrubs or replace them with shorter plants.
- Be sure the ground around the house slopes away from it and that no dirt touches any section of wood.

planted close to the house can add to the problem. Some quick work with pruning shears to cut back the plants and spot scrubbing on the siding should help control the problem.

Exterior mold starts slow but can grow if left alone. It's more than an unsightly problem; it can destroy paint and, unless it's cleaned off and eliminated, it will affect future repainting as well. Catching mold early and keeping it under control can easily be part of normal yearly maintenance.

YELLOW LIGHT

Pressure washing will remove surface mold, but you can't depend on it to do a complete job. The mold oftentimes has to be scrubbed to remove it. To do so with a pressure washer alone would likely require so much pressure the siding or finish could be damaged.

Deck Sweeping

- Keep decks regularly swept so spores can't get a foothold.
- Wash decks regularly with a mild soap and water solution—a regular garden hose will do; a pressure washer isn't necessary for maintenance cleanings.
- Seal the deck once a year before the weather turns cold with an exterior stain or clear coat containing fungicide.
- If a deck is in poor but restorable condition, consider sanding the boards smooth so they're less likely to trap and hold spores and water.

Siding Cleaner

- To clean exterior mold from walls, mix a solution of disinfectant cleaner, 1 quart of laundry bleach, and enough hot water to fill a 1-gallon bucket.
- Scrub the affected surfaces with a soft scrub brush, avoiding nearby plants.
- Rinse with a garden hose, again avoiding any plants.
- Repair any damaged paint areas by standard scraping, priming, and applying two coats of new paint with added fungicide.

SERIOUS MOLD

Regularly inspect your home for mold or mold-friendly conditions to avoid unwanted health problems

In a home without ventilation or moisture problems, any mold that shows up is most likely your standard unsightly annoyance. Some people, though, are sensitive to molds. According to the Environmental Protection Agency, "For these people, exposure to molds can cause symptoms such as nasal stuffiness, eye irritation, wheezing, or skin irritation. Some people, such as those with serious allergies to molds, may have more severe reactions. Severe reactions may include fever and shortness of breath. Some people with chronic lung illnesses, such as obstructive lung disease, may develop mold infections in their lungs."

That's the scary talk. Unless you have had flooding, ex-

Serious Mold

- Serious mold results from ongoing water leakage or seepage problems that go unnoticed or unattended to.
- Sure signs are softened drywall, soggy floors, and large affected areas.
- Extensive mold can be a health issue for people with allergies, asthma, or other lung diseases and can result in symptoms from a skin rash to difficulty breathing and nausea, depending on the mold.
- If you even consider removing damaged areas yourself, wear protective clothing, gloves, and a respirator.

Black Mold

- So-called black (toxic) mold produces hazardous microtoxins and can be a health risk.
- Black mold, as referred to in the media, is one of the *Stachybotrys* types of mold, shown to be harmful to humans.
- These molds grow in wet, humid conditions and are eliminated when the wet conditions are eliminated.
- Blue, black, or green stains on ceilings and walls are signs of black mold, as are brown, orange, or green stains on tiles and grout.

tremely high humidity, a leaking roof, or other moisture problems, the chances of your home harboring toxic mold are slim. Vigilance is still important, and regularly inspecting your premises for any mold-friendly conditions should be on your list of things to do. If you can smell mold—it will be an earthy or musty smell—explore for the source. If you discover a major source of mold, walk away and call a remediation contractor who is equipped to properly remove and clean the affected area. After that, moisture control will be the key to preventing mold's return.

ZOOM

The main culprit when referring to toxic mold is *Stachybotrys chartarum*, a fungus that produces toxins harmful to humans. It became known as "black mold" in the 1990s. It is a concern in flooded homes and is not a common occurrence. See the Resources section for more information.

Mold on Manufactured Siding

- Manufactured siding is composed of wood chips or flakes, wax, resin, and other adhesives that can deteriorate and cause mold growth, including mushroom growth.
- Not all manufactured or composite siding goes bad; it depends on the installation and maintenance.
- Once mold growth spreads on this siding, it's unlikely it can be successfully cleaned and the spread halted.
- Individual boards can be replaced and the others sealed and repainted to salvage less-damaged siding installations.

Hiring a Professional

- Removing and decontaminating extensive mold aren't a do-it-yourself activity.
- With widespread mold, hire a consultant to evaluate the problem and a separate, unrelated contractor to do removal.
- Hire an experienced, bonded, insured, and licensed contractor and check all references with the Better Business Bureau and your state's labor and industries department.
- Be sure the cleaning and abatement process and costs are clearly spelled out in advance as well as any warranties that apply.

ROOF MOSS

Moss can deteriorate your shingles, so kill the moss to preserve your roof

Moss is a survivor. It doesn't come with conventional roots, grows only in moist locations, and doesn't require dirt to grow, so it plants itself in all kinds of crevices and barren spots. Moss starts as wind-driven spores and, in damper climates, ends up on the roof as roof shingles stay wet long enough for it to get a foothold and grow. It normally develops on the more shaded side of the roof, which does not dry out as fast as the sunny side. The same is true for its growth on trees: In northern latitudes, it mostly grows on the north side of trees.

Moss may be a welcome addition to some gardens and between paving stones, but it's not welcome on a roof, par-

Roof Moss

- There are two main ways to remove roof moss: gentle scraping and pressure washing.
- Scraping calls for a light touch with a putty knife. It's tedious but effective and doesn't require renting a compressor and sprayer.
- After removing the moss, add a diluted solution of water and bleach to a garden sprayer to kill any remaining moss.
- Unless physically removed from the roof, dead moss can retain moisture on the edges of the shingles, attract new spores, and ultimately grow new moss.

Scraping Shingles

- Composite shingles are durable but vulnerable at the edges if you scrape them too aggressively.
- Lightly scrape off any moss with a narrow putty knife, being careful not to scrape granules off nearby shingles.
- Be cautious moving around on your roof—you don't want to slip or damage any shingles as you walk across them.
- After scraping, blow off the loose moss with a garden blower, being sure to clean out the gutters at the same time.

ticularly one with composite or fiberglass shingles. Even if the moss dies out over the summer months, the dead moss holds water once the rainy season starts and deteriorates your shingles, as they will stay wet, which loosens the mineral granules embedded in the shingles. Kill the moss and help preserve your roof.

ZOOM

Use extreme caution when cleaning a composite roof with a pressure washer. It's easy to damage the shingles and create bigger problems than the presence of moss. Never shoot up at shingles, always down, and use a low pressure. Pressure washing roofs is a controversial subject. Consider less aggressive techniques if at all possible.

Pressure Washing

- If you hire out a pressure washing job, be sure your contractor does this work as a specialty, not as a sideline.
- Follow up the pressure washing by spraying the roof with a mild bleach solution or other biocide to kill any remaining moss.
- Trying to blast the debris out of your gutters with the pressure washer can fill the downspouts with moss and granules from the shingles—be sure to block off the top of the downspouts with leaf guards.

Bleach

- Bleaching a roof with ordinary laundry bleach to kill and control moss is another practice that isn't universally recommended.
- Pouring straight or partially diluted laundry bleach on composite shingles might cause damage and shorten their working life.
- Home and garden stores sell commercial moss retardants based on zinc compounds that are sprayed on without damage to roofing shingles.
- Any moss killer can affect garden plants and calls for careful application following the manufacturer's instructions.

WOOD ROT

Rotting wood isn't only gross—it's unsafe and must be repaired right away

The term *rot* is tossed around too casually. Severely weathered wood that's rough and splintered after years of weather exposure is not rotted wood. Wet wood is not automatically rotted. Rather, soft, "punky" wood that pulverizes into powder or disintegrates when poked with a screwdriver is considered rotted. Another misused term is *dry rot*, which is really the observed results of rot once water and moisture problems have been eliminated.

Rot results when certain fungi find wet wood to dine on in a damp, preferably dark environment. Fungi are selective diners when it comes to wood and prefer the cellulose or lignin—parts of wood's cell structure—depending on the

Rotted Wood

- Wood rot is found in dark, moist areas, such as crawl spaces, and anywhere dirt doesn't drain well.
- Remove all damaged wood, bleach or apply another fungicide to the rotted area, allow it to dry, and replace the damaged wood with either epoxy or with new wood.
- Use caution around rotted structural lumber.
- Repairing structural framing calls for a carpenter's expertise and possibly a building department permit.

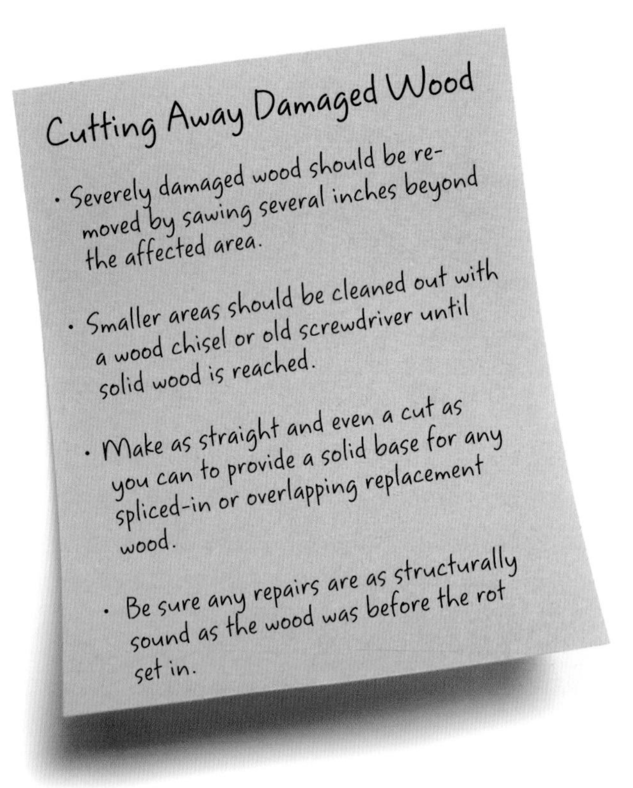

fungus. As they chomp away, the cell walls weaken and eventually collapse.

Rot manifests itself with white or brown staining and very soft wood. The worse the rot, the more damaged the wood. The first step when repairing is to stop the water problem. The rot will return as long as wet conditions allow it. The wood should then be replaced or treated and repaired with epoxy filler. Extensive rot that has infested a large section of a house foundation calls for professional repairs.

MAKE IT EASY

You might not have rot now in a crawl space or in the enclosed area under a porch, but why not head off future problems? Check that gutters and downspouts are intact and not leaking into these areas. Redo any landscaping graded towards the house. Inspect pipes for leaks or corroded sections that need monitoring.

Spraying Bleach

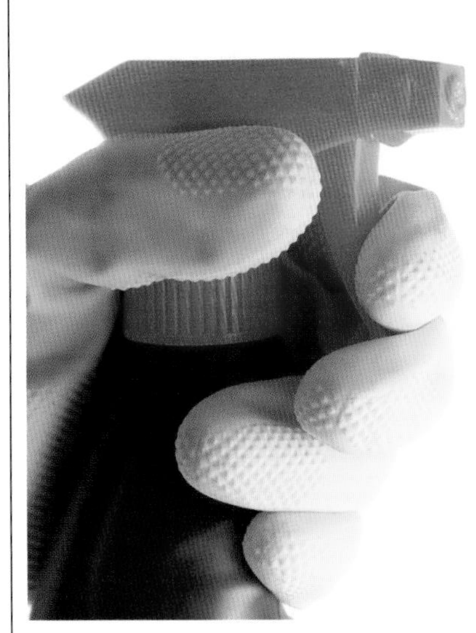

- After removing all rotted wood, spray or apply chlorine bleach (wear gloves and eye protection) or a commercial fungicide liberally over the damaged areas.
- Allow the area to completely dry before continuing the repairs.
- Drying can be sped up with a heat gun.
- As with any water-related repair, be sure you've eliminated the leakage or exposure that caused the rot in the first place in addition to repairing the damage.

Epoxy Repair

- Small areas of rot can be repaired with any number of epoxy products available at paint and marine supply stores.
- Epoxies set up very quickly—fully prepare the area before mixing the epoxy ingredients.
- Build up the epoxy in layers according to the package instructions.
- Sand or file the repaired area smooth if it's finished wood and seal with primer and paint; otherwise, concealed areas can be left with a rougher appearance.

INSTALLING A PET DOOR

A properly installed pet door allows your pets to come in but keeps weather out

A pet door provides an opportunity for your pet to enjoy a more natural environment and, at least with dogs, eliminates the need to "hold it" all day until someone comes home. Pet doors are convenient for both the pet and the owner and, when installed properly, minimize cold air from entering a warm home.

Pet doors match pet sizes and come in a wide variety of styles. Some are manually operated—the pet pushes it open—and others have electronic locks on them that respond to a sensor on the pet's collar. A pet can come and go at will with either type of door. A pet door can be installed through an exterior door or through an exterior wall.

Basic Facts about Pet Doors

- Purchase a door for the type of installation you're doing: door-mounted or wall-mounted and always read the instructions first.
- Size the door to your pet and mount it at a comfortable height for your pet's use.
- Medium and large pet doors are a security risk for someone who is small enough to gain entry or who reaches through with a stiff wire device to unlock a nearby door.
- Large pet doors might require a wall stud be removed and as well as some framing work.

Wall Installation

- With a stud finder, check that a wall stud isn't located where you want to install the door.
- Level the template that comes with the door on the wall and use it to locate and mark its four corners.
- Pencil straight lines among the corners and score (if plaster) or cut through (if drywall) with a utility knife.
- Push in and remove cut drywall; drill a hole in each plaster corner and cut the plaster out with a keyhole saw or a jigsaw.

A wall installation provides more options and can more easily be filled in later if the pet door is ever removed. Properly installed, a pet door will be sealed to the weather, benefit your pet, and buy you some time if you're ever running late returning home.

YELLOW LIGHT

Pets take to pet doors at their own pace. Don't expect yours to instantly master the door. Some take, seemingly, forever, but there are strategies to help them along. Don't assume you can neglect dog-walking duties the same day the door goes in—they'll adapt to it when they're ready.

Wall Installation (Continued)

- Cut away and remove any wall insulation.
- Carefully line up a long drill bit with the four corners of the cut-out area and drill through the exterior wall.
- Mark the outside wall between the drilled holes, checking with the template, and cut the siding and sub-siding away using a jigsaw or reciprocating saw.
- Trim out the installed door with wood molding and caulk as needed, especially on the outside for weather protection.

Door Installation

- Installing in a door is similar to wall installation except there's no chance of running into a wall stud, wiring, or plumbing.
- Remove the entry door for easier cutting and installation.
- If a paneled door is used, the gap between the pet door frame and the panel might have to be filled in with narrow strips of wood.
- Caulk around the edge of the cut-out area and press the pet door into the caulk to assure a good seal and to protect the wood.

SQUIRREL & RACCOON DAMAGE

Some animal damage you can tackle on your own, and some needs professional help

Squirrels and raccoons, like any animal, want to survive. They snoop around for a good place to set up house, and if it's your house, they're usually willing to share it with you, but you don't want to be sharing with them. Squirrels and raccoons use insulated attics as their own private getaway while leaving a path of destruction in their wake.

Aside from the chewing and clawing away at siding and insulation, squirrels and raccoons can damage wiring and any items stored in the rooms they infiltrate and leave a fetid, unhealthy mess of urine and feces that will damage the surrounding areas. They also reproduce. Depending on your state laws, it may or may not be legal to kill these trespassing

Basic Steps

- Watch for tell-tale signs of possible pest damage including birds flying around roof rafters, holes in siding, rats/mice around foundations, or any rustling or live sounds from inside walls or attics.
- These critters must be out of your house before you close up their entrances.
- After removal and complete cleaning and abatement, check for wood and wiring damage.
- Once they are removed, continue to do regular checks for return visits, both in the original location and new ones.

Removing the Damage

- All nesting materials should be removed and the area sanitized with bleach or other disinfectant.
- Wear gloves, a disposable protective suit, and a respirator—animal droppings are a health hazard, particularly if they've dried and become airborne.
- After cleaning and the area is dry, replace any damaged insulation with new material.
- Any wiring damaged by squirrels from chewing must be repaired—call a licensed electrician for replacement.

rodents, and it's doubtful you want to do it yourself in any event. Hire a professional exterminator and then deal with the clean-up, repairs, and preventative measures that will keep other squirrels from finding their way into your home.

Also, never try to handle a live animal, particularly if the animal is acting strange or is active at an odd hour (a raccoon walking around during the day, for example). Call animal control to deal with these pesky critters.

RED LIGHT

Never seal up access holes used by rodents, birds, bees, or any other animals or flying insects that have gotten into your house until you're sure they have all been removed. Trapping them inside without an escape route will make matters only much worse. A professional exterminator will guarantee the job—call one.

Spraying Insecticide and Fungicide

- Spray the entire area with insecticide to kill off any bugs or parasites that were living with the wildlife.
- Carefully follow the insecticide instructions and be sure it's appropriate for parasitic bugs.
- Limit your time in enclosed attic spaces when spraying insecticide and afterward or consider having an exterminator do the job.
- A second application might be called for, depending on the extent of the infestation and the chosen insecticide.

Calling a Professional

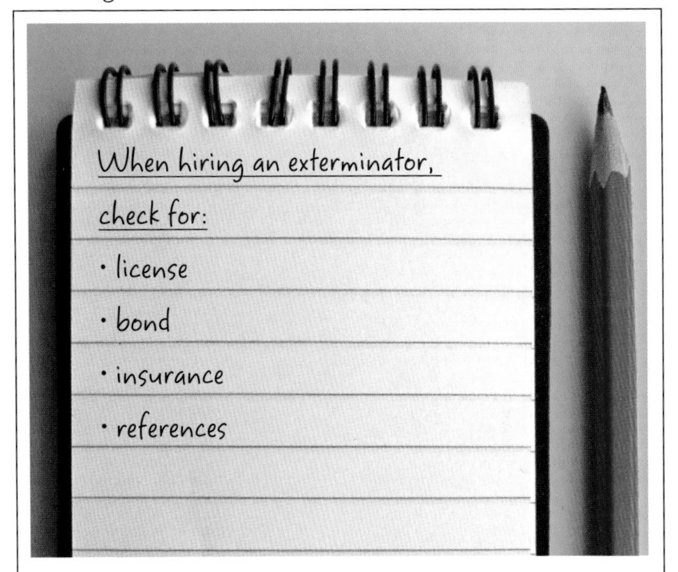

- It's one thing to remove a small amount of damage, quite another to remove large areas.
- If the infestation is extensive, call a specialist in pest removal and restoration.
- Some services will exterminate as well as do restoration work, repairs, and rodent prevention control all in one package.
- As with any contractor, check for license, bond, and insurance as well as references.

BIRD DAMAGE & REPAIRS

Birds can damage more than just your property—they can also spread disease—so remove them quickly

Birds, like rodents, will live wherever they can find acceptable shelter. Oftentimes, this means using your home for their nesting grounds. An attic or area protected by a roof overhang makes a great shelter: It's relatively dry, concealed, and close to food. However, their nesting habits leave you with all the clean-up.

Birds also do considerable damage from feces and urine, which can transmit diseases to humans from both physical contact and inhalation. Birds reproduce quickly, so you end up with two generations if you don't get rid of them early. You might need to call an exterminator if you find yourself up against particularly stubborn birds. Remem-

Birds in Attic

- Birds and bats will contaminate attics and any nesting places with urine and feces, both of which are health hazards.
- Pigeons or other birds roosting in covered roof areas will stay away while you remove a nest but can return later unless the area is blocked off.
- Something as simple as playing loud music in an attic space can drive adult birds out, but be sure no offspring are left behind.
- Rubber owls and snakes offer only temporary relief.

Bird Removal

- Once the birds and nesting are removed, thoroughly clean the area with an all-purpose disinfectant cleaner and bleach solution.
- Any shingles that are damaged due to bird wastes should be replaced and any damaged painted areas cleaned, sanded, primed, and repainted.
- After removing the birds and repairing the area, monitor it for their return.
- Check other areas similar to the one they were using and block these off as well before the birds relocate.

ber, they are not your friends—don't let them nest on or in your home.

If there is extensive damage, leave the clean-up to a professional, who will take away any hazardous material.

MAKE IT EASY

Any time you see the beginning of a bird nest near your roof, remove it immediately. Emotions set in when you see a nesting pair of birds and then their offspring, and the tendency is to leave them until the young are old enough to survive. Meanwhile, they're damaging your house. Stop it before they get started.

Cleaning the Mess

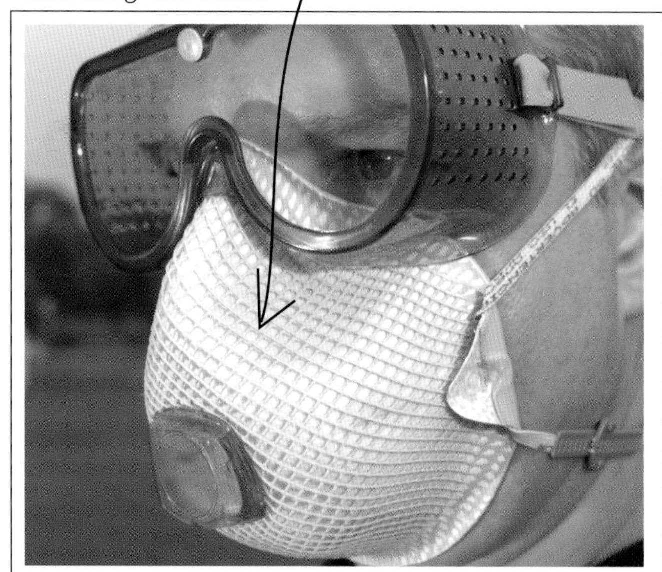

- Bird droppings are dangerous in both wet and dried form, but especially dry because they can pulverize during removal and become airborne and inhaled.
- Protective gear including gloves, disposable coveralls, and a respirator with HEPA filters should be worn.
- All contaminated debris should be double-bagged and disposed.
- Extensive damage should be left to professional wildlife and pest removal experts who will vacuum out all hazardous materials and clean the structure.

Installing Rails, Bars, and Spikes

- Sealing every possible access point should keep birds and other animals out.
- A simple solution to prevent birds from returning to roosting areas under roof overhangs or porch ceilings is installing a series of wood rails or bars, spaced so birds cannot pass through.
- Another method is installing stainless steel bird spikes, which are attached with adhesive.
- Do not nail any bird controls into the roofing shingles—attach them to the siding or wood trim.

BEES & WASPS

It isn't enough to eliminate these insects; you have to eliminate what they leave behind, too

According to Harvard University's Dr. E. O. Wilson, there are nearly 10 quintillion insects in the world. We will never live without them nor would we want to, but you don't want to live with them buzzing inside your walls or attic or even too close to your lawn furniture. Once again, your home makes a convenient and sheltered living area for flying squatters such as bees and wasps. All they need is a small opening through the siding to get started.

Honeybees will not do structural damage, but they will leave honey and wax combs that can attract rodents and other insects that can do damage to a wood structure. It isn't enough to remove or kill the bees; you have to get inside the

Wasps

- Locate active wasp nests during the day, but spray at night when the nest is full.
- Wasps can often be eliminated with over-the-counter aerosol insecticides.
- Yellow jacket nests last one season—if they're not particularly bothersome, but in an awkward spot to remove, you might decide to leave them alone and remove the empty nest in cold weather.
- After spraying an outside nest and confirming no wasps have returned, remove with a long pole and rinse with a hose.

Insecticide

- Like any poison, insecticides must be handled carefully and used effectively.
- Read and follow label directions for your chosen insecticide and always spray from a safe distance, being sure no other people or pets are nearby.
- Always store insecticides safely where children cannot access them.
- Some sprays have an oil base and leave a residue that should be washed off any painted surfaces the next day.

wall and remove any honey and deodorize the area. As with other home invaders, you don't want to seal off the entry point until all the insects have been removed.

Yellow jackets, a type of wasp, don't leave honey, but they do sting and can chew through drywall. If you can live with their presence until winter comes, freezing temperatures should kill them off and provide you with time to remove old nests, which will not be reused by the insects in any event.

RED ● LIGHT

Wasps are aggressive. You don't want to approach a nest during the day to exterminate it. Wait until evening when the nest is full before spraying with an appropriate insecticide. Use a projectile spray for elevated nests. These sprays will shoot 15–20 feet or so, keeping you a safe distance away.

Trap

- Several disposable and reusable yellow jacket traps are available that attract yellow jackets away from your home for easy removal.
- Directions for homemade traps are available on the Internet when searching under "yellow jacket traps."
- Wasp traps are generally nonpoisonous and make disposing of dead wasps simple.
- Traps cannot quickly counter an entire nest worth of wasps but can be effective for nuisance wasps in your yard or picnic area.

Bees in the Wall

- Destroying an established honeybee colony is not a do-it-yourself job.
- The bees must be removed or killed without being trapped inside your home.
- Call an experienced exterminator, some of whom might include removal of any wax combs and honey created by the bees.
- To do this removal and clean-up yourself, you will mostly likely have to remove sections of a wall. Complete removal is critical to prevent damage from fermenting honey that can attract other insects.

TERMITES & WOOD EATERS

These insects not only move in, but also eat your home out from under you

Termites are the most destructive wood-destroying insects in the country, causing over $2 billion in damage each year. In nature, termites are helpful. They break down organic matter, which then rebuilds soil and starts the plant life cycle all over again. In your home, they munch on framing lumber, window sills, trim boards, anything that suits them.

Termite infestation is considered more of a problem in the southern states, but termites can be found in every state in the country except Alaska.

Termites survive in warm, moist soil and travel in thinly constructed mud tunnels to their food source. Any wood with ground contact is an especially inviting target. Wet founda-

Different Insects

- Wood-destroying insects include termites, various wood-boring beetles, and carpenter bees, which excavate wood to build nests.
- Each cellulose-loving insect requires a different extermination approach and insecticides, some of which are best left to a professional service.
- Carpenter ants will nest in wet wood and can leave shredded wood near their nests.
- Homeowner-applied insecticide can eliminate the ants if all nests are found. More than one application might be needed.

Termite Damage

- Signs of termites include pencil-thin mudlike tunnels (termite tubes) along a house foundation, cellar walls, wood posts, and exterior wood trim.
- Because termites destroy and hollow out wood from the inside, damages go unseen for years.
- Exterminators inject an insecticide around a house specifically for termites, exposing them to poison as they return to their nest.
- In extreme cases, whole house eradication calls for tenting a residence and either killing the termites with fumigation or extreme heat.

tions also attract termites, as does any wood—firewood or lumber—stacked too close to a house.

Termites have some competition for your wood: Carpenter ants, powder post beetles, and other borer-type insects all take their toll and require professional extermination. Consider preventative treatment through a reliable local exterminator, who is an expert in handling poisons, before the destruction starts. Although you should not attempt to rid your home of termite and woodeater infestation on your own, you can recognize termite clues and eliminate what attracts them.

Repairing Damage

- Once the insects are eliminated, repair the damaged wood with epoxy or cut out damaged sections and replace with new wood.
- If the wood was sprayed with insecticide, spray the area with a bleach solution first and allow it to dry.
- Contact the manufacturer of your insecticide for information on a safe timeline for working around sprayed areas.
- Extensive foundation damage might require a permit and inspection before repairs can be started.

RED LIGHT

Masonry buildings are not immune to insect damage. Termites and other wood-boring insects will find any wood source as long as conditions are attractive. The wood framing behind a brick or stone facade will be the target, as will cardboard or paper, including the paper covering on drywall.

Correct Moisture Problems

- Correct all moisture problems, including roof and plumbing leaks, to prevent wood-destroying insects from establishing themselves.
- Be sure that there is no standing water near your house and that the ground is graded away from the foundation.
- Relocate any firewood away from your home and stack it off the ground.
- Seal foundation cracks and openings around pipes entering the foundation, and be sure there is no wood-ground contact of any kind other than treated lumber rated for ground contact.

CRITTER-PROOFING YOUR HOME

Keeping pests out early is easier than trying to evict them later

An ounce of prevention may be worth a pound of cure, but when it comes to keeping animals out of your house, a few boards and some screening can save you a bundle of money not spent in cleaning out and repairing a damaged attic or wall space. If you haven't had any four-footed or winged visitors, the first step is to do a complete exterior inspection.

Look for any hole, opening, or gap—for instance, between a pipe and the surrounding concrete foundation. Rats and mice can squeeze into impossibly small holes. Pay attention to any wood that appears to have been gnawed at or is rotted. Holes should be filled and compromised wood should be repaired. The goal is to completely seal off your house.

Installing a Chimney Cap

- Chimney caps protect the inside of the chimney from rain and animals while allowing the normal chimney functions to continue as long as the cap isn't too close to the top of the chimney.
- The heavy screening on the cap's four sides is sufficient to keep critters out.
- For easier chimney cleaning, look for a cap with a hinged top.
- Find the cap you want and inquire with the manufacturer for the needed chimney measurements before purchasing.

Installing Screening

- Use premade lattice to block off roof sections from raccoons and birds.
- Install 1/4-inch 16–19-gauge hardware cloth across louvered openings and to fill gaps between the exterior walls and the ground, burying the hardware cloth at least 2 inches deep.
- Steel wool and aerosol spray foam work only as temporary fillers in holes and gaps.
- Check for gaps between the bottom of garage, basement, and utility room side doors and install thresholds or weather stripping sweeps (see page 172) if needed.

Birds will roost under small overhangs if they have a roof or other support to rest on. Any probable landing spot should be closed off with lattice or screen. Water drainage is critical. You don't want soggy areas next to foundations or standing water under downspouts. Anything that attracts animal or insect life is fair game, including restricting access to food sources. A compost pile is a fine idea, but an open one is an invitation, particularly if you include food scraps in the mix.

MAKE IT EASY

Sealing off your home is an inexpensive proposition. For the price of some screen, lattice, caulk, and concrete patch, you can eliminate many if not all of the possible entry points for unwanted animals and insects. The key is being diligent and remembering that any opening presents an opportunity, so attend to them all.

Patching Concrete Foundation

- Seal up all openings with suitable materials: holes in concrete foundations filled with concrete patch, holes in siding filled with epoxy filler or new siding, and so on.
- If appearances aren't too important, use sheet metal to cover large gaps in siding.
- Rats and mice can squeeze into very small spaces around pipes and conduit passing through outer walls—be sure to fill and seal these.
- Seal any gaps between roof vents and clay tiles.

Securing Trash Cans

- Remove food sources that attract insects or other pests, including firewood, trash not secured in trash cans, pet feeding bowls, and bird feeders.
- Add a rag soaked in ammonia inside a full trash can to discourage raccoons from getting into the cans.
- Leaving food out for any form of wildlife will also attract animals you don't want to feed.
- Collect and dispose of fallen fruit and vegetables from gardens if rodents are a problem.

PICKING THE PAINT & SHEEN

Anyone can paint, but first know what you're painting with

Walk into any paint store, and you'll be confronted with cans and more cans. They all look very similar until you start reading the labels: high-gloss enamel, low-luster acrylic wall paint, flat matte emulsion, and alkyd modified latex. It can be confusing when all you want is to paint a bedroom. So let's walk through what you'll need. Basically, consumer paints are either water-based (latex) or oil-based (alkyd). Oil-based paint is rarely used now, although there are applications it's quite suited for.

For painting that bedroom, latex paint is easier to apply than oil-based, dries faster, is more forgiving during the application, and has far less odor than the solvent in oil-based

Bathroom

- Flat paints absorb light so fewer imperfections show, give rooms a softer feel, have less resin than glossier paints, and do not clean especially well.
- Ceilings and areas of least wear and tear are good candidates for flat paints.
- Satin and eggshell are low-luster paints with moderate gloss, making them easier to clean than flat paints.
- Paint these finishes in bathrooms, kitchens, children's rooms, and hallways, anyplace that will get a lot of hands on the walls.

Living Room

- Flat paints are easier on the eye, especially in well-lit rooms.
- Semigloss enamel has more resin than low-luster paints and stands up well to wear and tear and cleanings.
- Semigloss and high-gloss paints are common choices for woodwork, doors, and cabinets in older homes and are used for recoating in these homes.
- Because of its reflective properties, semigloss paint highlights a surface's imperfections, so thorough preparation work is a must before applying this finish.

paint. Paint is composed of three main ingredients: pigment that gives it color, a binder or resin that forms the film, and a vehicle that keeps the paint in a liquid form. In latex paint, water is the vehicle. As the water evaporates the paint film forms to a solid finish. Other ingredients are added by each manufacturer for different purposes such as to provide stability or to regulate the drying time. A higher gloss has more binder in the paint. More pigment generally means a lower gloss and better hiding power.

MAKE IT EASY

If a surface needs regular washing—doorways, for instance, because of handprints—don't use a flat paint. High-gloss paint will show every imperfection, so consider this before applying. If you're uncertain about a gloss choice, paint a sample first with a pint of paint before ordering it in gallons.

Interior Door

- High-gloss paint is the most durable and hardest paint as it has the highest percentage of resin.
- High-gloss oil-based paint demands more skill from the painter applying it, has a longer drying time than latex paint, and has more long-lasting odor.
- Experiment with higher-gloss finish (in a small bathroom, for instance) to test out the shine.
- You can always recoat a new finish with a gloss that is more suitable.

Summary of Finishes

- Flat: Apply on most ceilings and walls except those needing regular cleaning; not appropriate for woodwork; less reflective, a good finish for less-than-perfect surfaces; different manufacturers have different degrees of flat—some have more sheen than others.
- Satin and Eggshell: low sheen, washable, use in bathrooms and kitchens or any surface that needs washing; also works on woodwork.
- Semigloss: shiny and tough, use on woodwork and furniture.
- High-Gloss: shiniest of all, use on woodwork and furniture.

INTERIOR PREPARATION

The best paint won't cover up a lack of preparation work

Repainting a room that badly needs it is gratifying—and the gratification is immediate. The contrast between a fresh, clean coat of lemon yellow and the dirty, scarred pine green can be startling. But bad prepping will still show through on a fresh paint job. No one wants to prep; do it anyway if you want the paint to last and get the best results.

Prepping ranges from filling nail holes to washing the woodwork to extensive sanding, depending on the condition of the surfaces and the results you're looking for. If you hate old brush marks, for instance, they will have to be sanded out until the surface is smooth. Cracks in the ceiling bother you? Repair them before you paint, or you'll re-

Washing Walls

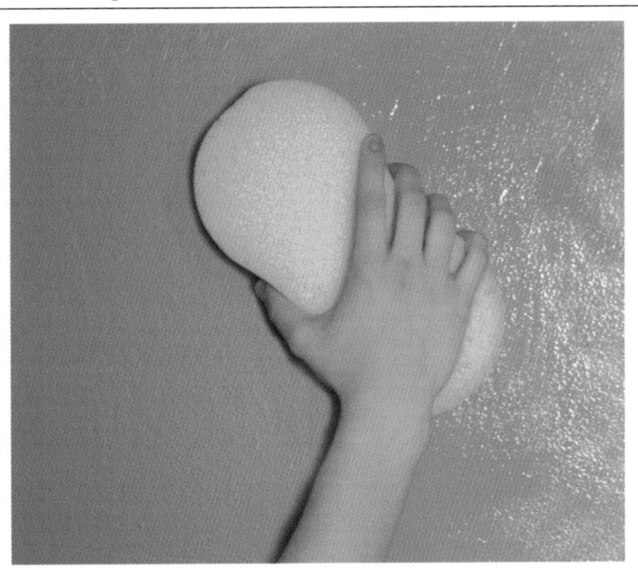

- Paint adheres best to clean surfaces.
- Woodwork should be washed with a nonsudsy cleaner to remove dirt, oil, and fingerprints.
- Wash bathroom and kitchen walls and ceilings with a nonsudsy household cleaner as well to remove all grime and grease.
- Although TSP (trisodium phosphate) has long been recommended for cleaning prior to painting, the phosphates in TSP can actually serve as a food source for mildew and promote its growth.

Lightly Sanding Woodwork

- Lightly hand-sand all woodwork to provide a better "tooth" for the next coat of paint and to smooth over nicks and gouges.
- Vacuum the dust and wipe the wood with a tack cloth, a waxy piece of cheesecloth available at paint stores.
- Dry sanding lead-based paint will release air-borne lead particles—wet sand only.
- Repair wall and ceiling cracks, caulk woodwork seams, prime bare spots and fill holes with Spackle or other filler, sand smooth, and prime.

gret it later. Kitchen and bathroom walls and ceilings get grimy even if you don't quite notice it with the passage of time. Wash and rinse them before painting. Brass hardware with paint drips on it can either get more paint drips or be stripped, cleaned, and polished, but you have to remove it first. It all takes time, and you won't regret a minute of it when you finish painting with superior results.

RED LIGHT

It's safe to assume any house built before 1950 has some lead-based paint in it. Even light sanding can spread contaminated dust. Wet sanding alone is impractical for removing deep brush marks. Special sanders with vacuum attachments plus containment procedures will be necessary for this type of prep.

Removing Hardware

- Remove as much window and door hardware as is practical.
- While the hardware is removed, it's an ideal time to clean, polish, and apply a clear coat of finish, especially to old, brass window and door hardware.
- With masking tape, tape off door hinges, door knobs, and other hardware that isn't removed.
- After removing light switch and receptacle covers, tape over the switches and receptacles.

Plastic/Drop Cloths

- Remove as much furniture as is practical and cover the rest and the floor with painter's plastic and tarps.
- Pile the remaining pieces of furniture on top of each other, leaving enough room to paint.
- Tape plastic around light fixtures, but keep the plastic away from the potentially hot light bulbs.
- Be sure to cover the entire area to avoid messy dripping paint.

PAINTING WALLS & CEILINGS

You can cover a lot of area fast with the right tools and a plan

You have a room to paint. Where do you start: the trim, the windows, the ceiling? There is some difference of opinion regarding painting the walls or the trim first, but most would agree to do the ceiling first. No matter how careful you are, there will be some drips and spatter from this, the highest section of a room. There was a time, before paint rollers were introduced, when all surfaces were painted with brushes off ladders and scaffolds. With today's tools, rolling out a room is much less tedious with more even results.

With a roller and extension pole, you can paint most walls and ceilings from the floor, using a stepladder to reach corners and cut-in work where the walls meet each other and

Painting Tips

- Fewer long strokes are better than more short ones.
- Keep the roller from getting too dried out to prevent it from spattering and applying paint in too thin of a coat.
- Wet the roller (or brush) as often as you need to for easy movement and full coverage.
- By sliding the roller cover slightly off the roller, you can cut in closer at ceiling and wall intersections, wall corners, and next to any woodwork.

Start with the Ceiling

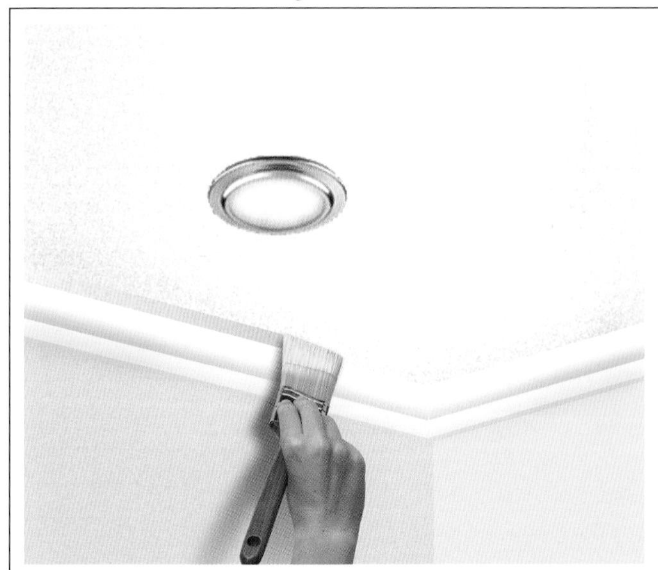

- Before rolling, paint a 2"X3" stripe around the edge where the ceiling meets the walls.
- Using a 3/8-inch or 1/2-inch roller and an extension pole, start at one corner of the ceiling and work along the shortest direction to the other side.
- Roll a 3–4 foot square "W" to distribute the paint and then roll in one direction.
- Glidden Ceiling Paint contains a pink dye that allows the painter to easily see any missed areas while painting; when dry, this paint turns white.

the ceiling or to paint around light fixtures. The texture of your surfaces or the degree of texture you want to impart with paint will determine the roller nap used. A stucco roller will cover unsightly areas on a relatively smooth wall that might have seen better days and has been repeatedly patched. The key to painting any wall or ceiling is to aim for even and complete coverage, using plenty of paint, and avoid roller marks. It isn't a race—take your time until it looks right.

YELLOW LIGHT

A spray-on textured ceiling that has never been painted can loosen when first painted with a roller. Some painters will only spray these ceilings, while others claim careful rolling will also work. If rolling, use a deep-nap roller without pressing excessively into the texture. Prime first and roll as close to the walls as possible.

Cutting In

- If the walls are a different color from the ceiling, allow the ceiling to dry and then do the walls, "cutting in" where the walls meet the ceiling with a paintbrush, in each corner where two walls meet, and around all woodwork.
- Work from one corner of the wall to the other, rolling horizontally at the top and bottom and finishing with vertical rolling, maintaining a wet edge from one painted section to the other.
- Apply enough paint for full coverage while avoiding sags.

Cleaning Up

- Use as much of the paint in your roller and brush(es) as you can before rinsing them out.
- Clean latex brushes and rollers with warm, soapy water and a brush/roller comb, rinsing repeatedly until the water runs clear.
- Shake out and spin to remove excess water—a roller spinner does this best—and allow covers to dry on paper towels; return brushes to their original cardboard brush keepers.
- Clean oil brushes with paint thinner and toss oil rollers out.

WOODWORK & TRIM

Some fancy brush work will make your woodwork and doors stand out

Even though there is less woodwork in most rooms than drywall or plaster, it can be more noticeable because of its glossier finish. Unless your woodwork is sprayed, which leaves a beautiful, smooth finish, yours will be brushed. It's important to use a quality brush, and that means higher-priced.

Skimping on brushes might affect your handiwork, since often the less-expensive brushes are also of poorer quality. For latex paints, use top-end nylon brushes, not blended nylon/polyester brushes. Oil-based finishes, both paint and varnishes, call for a China bristle or badger hair brush. The one-type-fits-all-finishes brush doesn't do a good job with any finish but only a passable one.

Brushing the Window Casing

- Paint woodwork after the walls and ceilings when brushes and rollers, rather than paint sprayers, are used.
- Dip the brush into the paint no more than halfway up the bristles, tapping both sides of the brush against the inside of the can to remove excess paint.
- Work from the highest horizontal section of woodwork down, painting in long, even strokes.
- Brush out any drips before the paint dries.

Painting the Window Sash

- Open the window sash—the movable part—any way necessary to completely paint it and the surrounding frame.
- New wood windows with integral weather stripping call for very careful painting to avoid getting paint on the weather stripping.
- If necessary, tape off the glass to avoid getting paint on it.
- If possible, leave each sash slightly open overnight—if any paint does hinder after drying, you can break the bond by moving the sash.

Although there's less square footage to woodwork, it's all cut-in work and almost always takes more time than the walls and ceiling. With a careful and steady hand, you can avoid taping off the wall where it meets the woodwork and just paint freely. You'll know after you paint your first door casing whether you prefer taping or going freehand. Follow whatever is comfortable for you to get the results you want. Keep in mind that taping is time consuming, perhaps more so than the time needed to paint slowly and carefully.

MAKE IT EASY

Paint operable wood windows or exterior doors early in the day so they can be dry enough to close at night if needed. Move sliding windows every hour or so to prevent them from drying shut in an open position and keep moving them until the time they have to be locked for the night.

Painting the Door

- When painting doors, paint panels first, then the horizontal sections of the door, and finally the vertical sections.
- The edge of the door that opens into a room should be painted the same color as that room's trim.
- Paint the top, hidden edge of the door as well—this helps prevent it from sticking later.
- If the door has multiple layers of paint, scrape or sand down the vertical edges to prevent it from sticking in the jamb.

Painting a Staircase

- Staircase railing parts include the newel posts, the railing, and the balusters ("spindles"); brushing them as a unit is a challenging job.
- Start on the vertical newel posts and balusters along with the bottom of the railing, getting the finish as even as possible.
- Next, paint the top and sides of the railing.
- If the railing is stained and varnished with painted balusters and newel posts, tape off each section before coating the other.

EXTERIOR PREPARATION

When paint has to withstand the weather, prep work is even more critical

Exterior paint preparation ranges from a simple wash and rinse with a hose to full-bore paint removal with sanders, heat, and chemicals. Newer homes with only the original paint coating on it still intact often need just enough scrubbing to remove surface dirt and pollution to get them ready for a recoat.

Older homes with multiple layers of paint can present multiple problems: flaking, blisters in the paint, checking, chipping, and cracking. Overflowing gutters can send so much water down the side of a house that entire sheets of paint will lift off. If a house has too many layers of paint, it becomes more difficult for a new coat to stick to the old.

Power Washing

- Power washing might require using a rented sprayer; be sure to clean off any mildew separately before power washing.
- A 1,200–1,500 PSI model pressure washer should safely wash wood siding.
- Start from the top down and spray carefully, keeping the sprayer far enough from the painted areas so they get cleaned, but the wood doesn't get damaged.
- This requires ladder work, which can be awkward at first due to the pressure of the sprayer. Be sure to use caution.

Scraping Paint

- Flaking, peeling, and bubbling paint should be scraped and sanded off as needed and all bare areas primed.
- Older homes have a good chance of having some layers of lead-based paint, so wear protection.
- Various government-documented precautions must be taken before removing lead-based paint—see Resources on p. 228 for more information.
- Paint remover can also be used, but the removal process is quite tedious.

Intercoat adhesion—the ability for one coat of paint to stick to another—is jeopardized by lack of preparation, so prepping is a must. Extensive prep work can be messy and time consuming, but it's extremely important not to skimp on this step—both on your own and with a contractor. The most expensive paint cannot disguise nor have a long life without the required prep work first. This is exterior painting—you want it to last!

MAKE IT EASY

As an alternative to power washing, use a car washing brush set, a telescoping or series of aluminum poles with a scrub brush attachment. Some kits have soap dispensers and attach to a hose, allowing both soapy and rinse water through the brush. Manually scrubbing the house surface can remove more dirt than power washing without the possibility of damage.

Caulking and Spackling

- Caulk any seams between trim and siding, and fill holes with Spackle or automotive body filler (see page 46), sanding smooth and priming when dry.
- Caulk window glass where needed, following the directions for curing times before painting.
- It isn't necessary to caulk narrow gaps under siding, which can cause problems by sealing the house too tightly.
- If spackling areas where paint has been scraped off to achieve a level appearance, prime first, Spackle, and prime again.

Color Selection

- Remember that your roof color and any masonry details will affect your color selections for the rest of the house.
- Dark and deep exterior colors are more prone to noticeable fading than light colors.
- Look at sample colors in average daylight conditions, not extreme sunlight or dark conditions.
- Instead of choosing a contrasting accent color, consider a lighter or darker shade of a color already in use.

EXTERIOR PAINTING

It's a big job, so break it up into smaller jobs

To make painting the exterior of your house more doable, break the whole project into sections and work on it during scheduled hours. You'll get it done while the weather is still warm and dry without taxing yourself. Most homes will have trim, soffits—the underside of the roof that extends over the siding—windows, doors, and siding to paint.

The siding takes up the most area but is the easiest to paint. Like with interiors, start at the top and work your way down. If you have a large home, you might want to get some painting bids instead of doing the job yourself.

How long will exterior paint last? It all depends on weather exposure, the condition of the wood, the quality of the paint,

Painting the Exterior

- Wide siding can first be rolled to distribute the paint quickly and then brushed.
- Brush in long, even strokes, working the paint up and into the siding edges.
- Never stop in the middle of a wall, always complete an entire length of siding, especially when using stain.
- Hang your paint bucket from an extension ladder using an S-shaped bucket hook—this is much more convenient than tying the bucket to the ladder or using wire.

Painting Shingles

- Shingles are an ideal candidate for spray painting—they require a lot of paint due to their rough surface and have a lot of spaces to fill between shingles—but they can also be rolled and brushed.
- Work the paint into all three exposed shingle edges, especially the bottom open grain, which really soaks up paint.
- Watch for drips between shingles—go back and catch them before moving too far down the wall.
- If your shingles are stained, they will absorb a lot of new finish.

and its application, all of which determine the life of a paint job. To improve the odds, use quality paint with a high percentage of solids. Apply during warm, dry weather and brush on a full, thick coat.

YELLOW LIGHT

Ladders can slip easily, so remember to fully secure any ladder, especially as you work higher up. It's a good idea to insert a screw eye in an inconspicuous but secure place—behind a barge rafter, which runs along the roof line, is good—for tying off the top of a ladder.

Be prepared to use a lot of paint for stucco.

Painting Stucco

- Stucco is usually painted with a coarse roller, one with a thicker nap, or first sprayed and then rolled.
- Prime any repaired stucco areas, including newly sealed cracks, before painting.
- Elastomeric paints, which are thick, protective rubberized finishes, are increasingly popular for residential stucco finishes and have the added benefit of filling in and sealing hairline cracks.
- Stucco's rough texture requires a lot of paint, but it rolls out faster than wood siding.

Weather Conditions for Painting

- Paint on a dry day, 50–85 degrees, and avoid direct sunlight.
- Follow the sun around the house so a freshly painted side is never exposed to excessive heat while drying.
- Follow the paint manufacturer's instructions for temperature and shade recommendations.
- Keep in mind that these will be the best, but not always realistic, conditions to try to assure the paint will not fail.

BUILDING SHELVES

Adding shelves multiplies your space and makes finding the things you need easy

Think of shelving as multiplying your floor space over and over again. Without shelving, everything would be on the floor, stacked one box on top of another, and horribly inconvenient to access. Put up a few shelves, and a mess of a garage becomes manageable, you're able to find your shoes in the closet again, and the kids no longer have an excuse for a messy room.

The beauty of shelving is there are virtually no limits to the styles, dimensions, and load-carrying ability of shelving. From the finest china teacup to an automobile engine block, there's a shelf that will hold it safely and securely.

Aside from the shelving material—wood, metal, plastic—the shelf supports run the range from strands of wire and

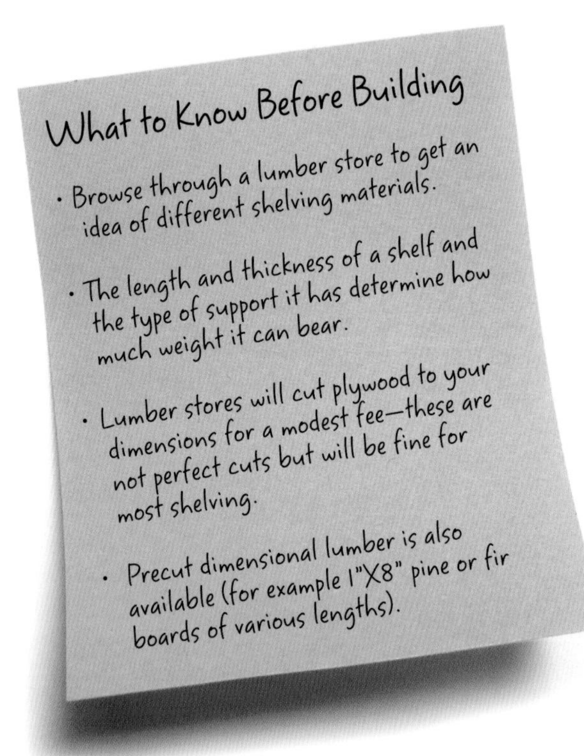

Studs and Supports

- Inexpensive metal utility L-shaped shelf brackets will support any length shelf as long as they are screwed into wall studs.
- The other side of the "L" screws into the shelf.
- Shelf support tracks, vertically installed steel channels called standards, screw into the studs and have adjustable brackets so the shelves can be installed in multiple locations.
- It's a simple carpentry job to make your own brackets out of short pieces of lumber.

cables with attaching hardware to 4"X4" posts. Hidden supports can also be used for a clean, floating look, and adjustable supports allow shelves to be moved and repositioned as storage needs change.

Shelving dimensions match the shelving material to the weight it must bear, so there are some design limitations, but they are few and far between, given the materials and designs available.

ZOOM

Plywood makes excellent shelving and is less prone to bowing than pressboard, particleboard, or oriented strand board, all of which are made from wood chips or fine sawdust mixed with different combinations of resin, glues, and binders. For either material, a 3/4-inch thickness should be considered minimum for most shelving.

Shelving Types

- Lightweight hanging shelves hold small items and hang from small nails or picture hooks.
- Built-in shelves call for removing wall studs and shouldn't be constructed in load-bearing walls without an engineering plan and proper support.
- Free-floating shelves have hidden supports for a clean, contemporary look yet are still strong enough to support moderate loads.
- Corner shelves make good use of unused space.

Sample Installation

- Your shelf should be at least 16 inches long and constructed from 3/4-inch to 6-inch-wide boards.
- Cut one ledger board for each shelf.
- Level each ledger board against the wall and attach with drywall screws into the studs; rest a shelf on each ledger and screw the shelf to the ledger, lining up the ends.
- Keeping the shelf ends level, cut two boards long enough to screw into the ends of all the shelves and reach the floor to provide support.

MORE SHELVING

Shelving kits, both preassembled and custom sized, are easy organizers

Shelving kits, some in the form of brackets and supports only and others with the shelves also included, have been around for decades. They make sense for many installations, allow maximum flexibility and are cost effective, but do not offer the finished look of fitted wood shelving.

Some finished, self-contained bookshelves offer the best of both worlds: a finished look with mounting strips or clips secured to the sides—where they're barely visible—for adjusting the location of the shelves. In some older homes from the 1950s and earlier, the builders included built-in shelving with nicely detailed wood cutouts for inserting shelves in various locations.

Wire shelf kits are a more contemporary feature in both

Shelving Materials

- Particleboard is available plain, with wood graining and a plastic coating (melamine), which doesn't require painting and is easily cleaned.
- Plywood comes in different grades, including finished on both sides, cabinet quality with different wood veneers (oak, mahogany, birch, etc.), and more common, rougher finishes.
- If 3/4-inch material isn't available, glue and screw narrower plywood/particleboard pieces together for increased strength.
- A 2"X12" board resists bending and requires fewer supports.

Channels and Brackets

- Steel channels or standards are available in different metal finishes and colors, lengths, and design and can be purchased by the piece.
- The brackets are not always interchangeable—be sure the brackets you purchase are made for your standards.
- Shelf standards are the simplest and most versatile attached shelving available, but they do not offer an especially finished appearance.
- The key to installing standards is that they are vertically straight as well as level with each other.

new homes and as individual installations. Vinyl-coated wire shelves with metal brackets are quite sturdy, but the gaps between the wires pose a few limits as to what the shelves can hold (they're not great for thin paper files, for instance).

On the other hand, wire shelves don't accumulate as much dust as solid shelves, nor do they hold food crumbs or other small bits of food that tend to show up inside pantries and kitchen cabinets. The quality of the shelves varies, like anything else, with the commercial, heavy-duty wire shelves being very good quality.

YELLOW LIGHT

Regardless of the type of shelves you install, the brackets or supports should be fastened into wall studs for the best support. If a stud simply isn't available, various anchors, including molly bolts and toggle bolts, are manufactured for securing objects to walls and ceilings when no framing is available.

Wire Shelving

- Wire shelving kits, complete with brackets, fasteners, and anchors, are sold in presized sets or by the foot, with installation and assembly hardware sold separately.
- Wire shelving is convenient and fills many shelving needs, but plywood shelving is more versatile.
- Anchors, including molly bolts and toggle bolts, can be used to secure low weight-bearing shelf supports to walls when no studs are available.
- If you need a lot of shelving, cutting your own wood assemblies can be more economical.

Unfinished Shelving

- The easiest finish for unfinished wood shelves is several coats of modified oil.
- If possible, finish wood shelves before installing them for ease of application.
- Particleboard shelves should be sealed with paint or polyurethane if there is any chance liquids will leak on them or if they'll be subject to regular cleaning.
- Extensive built-in shelving can be painted more efficiently with a spray gun.

CLOSET FIXES

Closets are big business, but you don't have to spend a fortune on yours

Closets come in all different shapes and sizes, ranging from small reach-in closets to room-sized walk-in closets. An entire multibillion-dollar industry has evolved around closet storage, expansion, and organization, but you don't have to spend a lot of money to revamp and repair yours.

Some closet makeovers are works of art with fine cabinetry and elegant lighting. Other closet makeovers are simply more thoughtfully placed shelves and storage bins to help maximize space. The size of your closet determines the volume of clothing you can store, and the design determines how much of that volume you can practically use. Installing clothing rods at different heights to hold clothes of different

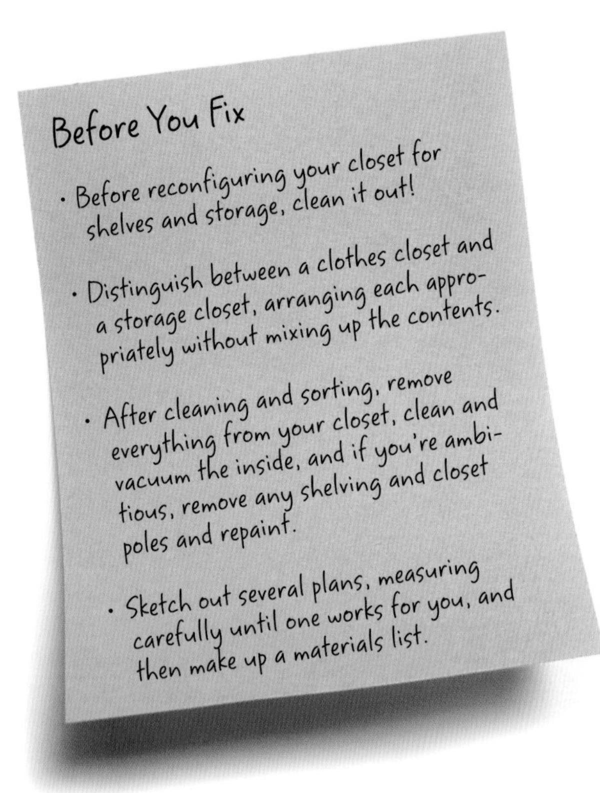

Additional Closet Rods

- Installing multiple rods at different heights makes better use of space.
- Closet rods require their own mounting hardware, both for rods that go from one wall to another and those requiring a bracket partway across the wall.
- Don't paint your closet rod; the finish will only get scratched and wear off every time a hanger slides across it.
- A steel pipe—not copper—cut to size with its own mounting hardware will resist bowing better than a wood closet rod.

hanging lengths can double the amount of storable clothes. Built-in drawers and shelves allow you to actually find a lot of items that end up on the floor. And extra shelving helps you store out-of-season clothing in an out-of-the-way location.

With these practical and inexpensive suggestions, you can turn your ordinary closet into an extraordinary space.

ZOOM

Consider tongue-and-groove aromatic cedar panels for paneling closet walls. Cedar is a natural insect repellent with a very agreeable scent. As the panels are exposed to air, the pores of the wood can close and restrict the cedar smell. Running very light sandpaper (220 or so) over the cedar will reactivate the cedar oil.

Professional Closet Upgrades

- Some businesses specialize in closet systems and build custom storage units that are beautiful yet expensive.
- Other companies provide do-it-yourself closet design tools and build and ship storage units of your design.
- An online search under "closet organizing" will bring up all the closet storage options you could ever imagine.
- Keep a budget in mind and stick to it—the price varies greatly depending on the type of closet, so do thorough research.

Shoe Storage

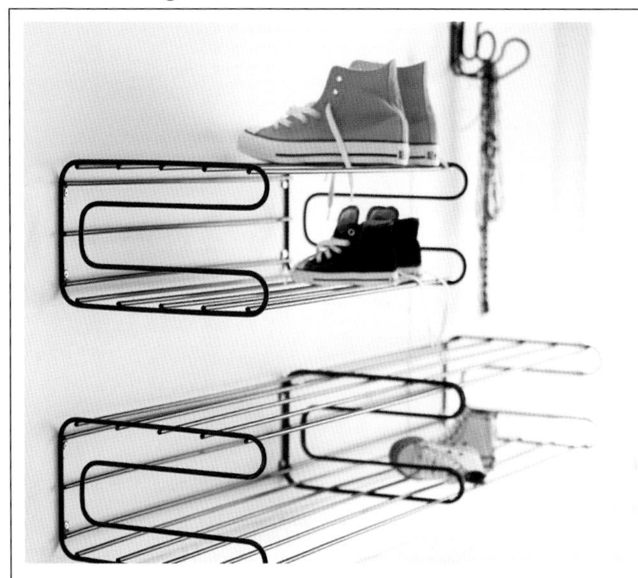

- Shoe racks range from small, adjustable shelves for a few pairs to multi-shelf shoe racks that hold dozens of shoes.
- Hanging shoe shelves are available for closet doors in both soft-style shoe holders and hard-wire style.
- Modular shoe racks that snap together allow for an expanding shoe wardrobe.
- The least expensive method is stacking the original shoe boxes with photos of the shoes on the side for better organization.

GARAGE STORAGE

With new shelving and storage systems, the days of the overflowing garage are gone

Garages were once utilitarian places. Cars, lawn mowers, bikes, garden tools, and boxes full of mysterious and forgotten contents packed garages in a first come, first stored manner. Now, entire companies are devoted to garage storage needs, including installations.

Garage cabinetry can rival that found in kitchens. Garage floor finishes have come a long way from bare concrete and latex paints that never quite held up under automobile traffic.

Upgrading your garage can be as expensive or modestly priced as you want. The main thing is to get as much as possible off the floor for convenient, dry storage and a safer garage environment (you won't trip on sports equipment if it

Modern Garage Storage

- Premade storage systems look great but are typically more expensive than do-it-yourself storage and are limited to the manufacturer's dimensions.
- Because garages aren't subject to as much scrutiny as living areas, you don't need to produce fancy storage.
- Building a custom-sized loft storage unit, supported by posts, could still allow a car to park underneath.
- In earthquake areas, items stored overhead must be secured to prevent slippage and falling.

Modern Overhead Storage

- Heavy-capacity shelving can be supported by steel hanging hardware fastened to ceiling joists.
- Pulley and hoist systems allow overhead storage of bulky objects such as kayaks, extension ladders, and large garden equipment with easy raising and lowering.
- Commercial overhead shelving is designed to fit above garage door tracks without interfering with the operation of the door.
- Overhead storage requires strong hardware and shelving—consider buying a commercial system instead of fabricating your own from wood.

isn't lying all over the floor). On the most basic level, a good cleaning and sorting goes a long way towards organizing a garage. Spray painting the walls and ceiling (whether they're finished with drywall or not) and then prepping and painting the floor will brighten up any garage for a low cost.

Plywood shelving can be precut at a lumber store to any desired width for custom shelving. Shelving can run up close to the ceiling to store those rarely used but must-keep items. With some planning and good execution, you might even be able to put your car in your garage again.

MAKE IT EASY

Consider how far you want to take your garage storage before you begin installing. Do you want a more finished look with painted shelves? Would cabinet doors be preferred over open shelving? Complete install-it-yourself systems are available at home centers. Check with a store employee to get the set kit you need.

Hanging Equipment

- Bicycles, sports equipment, and garden tools can be hung on walls or a ceiling with hooks or other hardware.
- Account for the people in the home so that equipment can be hung up and removed by children or shorter people, who would be better served by wall hooks.
- If bikes are hung on a wall, attach strips of carpet first so the bike tires don't mar painted surfaces.
- If you need more readily accessible equipment, freestanding racks rest against a wall, requiring no fasteners.

Plywood Shelving

- An 8'X4' plywood sheet can be cut into two 8'X2' shelves, plenty for most bulky items.
- Support these shelves with 2"X4" wall-mounted ledgers and 2"X4" horizontal and vertical side supports.
- The vertical supports will rest on a concrete floor where moisture can be present—use treated lumber or coat the ends of untreated lumber with wood preservative.
- When in doubt, add more supports.

WORKBENCH

Build a workbench for additional storage and for a clean, professional look to your work space

Every home needs a workbench. Where else can you stash your tools, disassemble a balking vacuum cleaner, or paint a chair you found at an estate sale for $2? A workbench doesn't have to be huge; even 4 feet in length will do, but it does have to be sturdy. Typically, a workbench is built out of framing lumber, including the top, so it can withstand hammering and general abuse. More elegant ones are finished off with scraps of hardwood flooring, but this is mostly for show.

A workbench can be elaborate with built-in drawers and cabinets or a basic four-legged affair with room for storage underneath. Metal industrial workbenches designed more for assembly work come with various arrangements of shelves

Figuring Workbench Dimensions

- The critical dimension to your workbench is its height—it has to be comfortable for you to drill, hammer, cut, and saw.
- Use your kitchen counter height as a comparison—if it's too high or low for workbench activity, adjust your height dimension accordingly before building a workbench.
- Make your workbench as wide as you can for extra project and storage space.
- Remember to figure in shelf, container, tool, and work space.

Simplest Workbench

- The simplest workbench is made from sawhorses and a sheet of plywood.
- Most preassembled wood sawhorses are too short for comfortable workbench activity—they're constructed in part to lay down plywood for cutting with a circular saw.
- Local lumber stores sometimes make their own sawhorses and can make you taller ones for a reasonable price.
- Folding sawhorses are appropriate for a workbench that will be used for light work; they fold up for easy storage when the job is finished.

and drawers as well as prewired receptacles that plug into a standard wall receptacle. Typically these are somewhat narrow but can be a bargain at surplus sales or going-out-of-business industrial plants if you need only a small workbench.

Building a workbench is a good exercise in beginning carpentry. The materials are inexpensive, so an incorrectly cut board isn't anything to worry about. You can even use old or scrap lumber, which keeps it out of a landfill and gives you something useful at the same time.

MAKE IT EASY

If you have the room, consider going longer rather than shorter with your workbench. It can always be used for storage, future projects, and multitasking.

Basic Corner Workbench

- Building a workbench in a corner of your workspace provides you with two walls for supporting the top and for hanging tools.
- Workbenches should be built near a 20-amp circuit GFCI receptacle and have plenty of light for safe working conditions.
- Construct your workbench with screws rather than nails if you're uncertain its location will be permanent.
- Exquisitely made workbenches can be ordered online—they're beautiful but expensive.

Workbench Tops

- The stoutness of the workbench really depends on its intended work load.
- Traditionally, workbenches have 2"X4" or 2"X6" tops to withstand heavy pounding.
- If you expect to do only lighter tasks, a plywood or even a particleboard top will work adequately.
- As time goes by and your workbench top gets too beaten up for your taste, it can always be covered again with another layer of plywood or 3/4-inch boards for a new appearance.

KITCHEN ISLAND

Whether movable or fixed in place, an island is a great kitchen addition

Modern kitchens are a marvel. They are larger than ever and contain every appliance imaginable. As the twentieth century developed, kitchens evolved from rooms solely for preparing food with separate dining rooms to the mixed-use gathering places they are today. We prepare food, eat it, and entertain all in the same room.

Newer kitchens are very open, with their own dining areas and usually adjoined to a family room. Older homes that have not had their floor plans modernized normally have closed kitchens with a doorway into a dining room. Either design can benefit from an island addition, but a closed kitchen must be large enough to accommodate an island without being cramped.

Benefits of a Kitchen Island

- A fixed island typically has storage under the countertop and built-in electrical receptacles, which run off a kitchen 20-amp circuit (see page 110).
- Although most fixed islands are built on a square or rectangular cabinet formation, the top can be shaped much differently.
- Movable islands eliminate space problems, and installing a floor-mounted receptacle provides power for appliances.
- A freestanding butcher block cutting station also serves as an island.

Premade Islands

- Multiple styles of movable kitchen islands on casters, both with and without storage, are available in a range of prices.
- For the ultimate in portability, folding kitchen work carts provide solid work surfaces and easy fold-away storage.
- Restaurant suppliers offer commercial stainless steel work stations as another alternative for an island.
- Custom-sized butcher block countertops are a food-friendly alternative for fixed islands that also act as cutting boards.

An island can contain a stove top and/or a range and a second sink, or it can be solely for storage with a countertop for food preparation and eating. You can build a simple island with a butcher block top or buy one of many portable islands, also called kitchen work carts, with a range of features and prices. The advantage of a wheeled island is portability to move it aside or move it to another residence. When you think your kitchen is maxed out with cabinets and counters, an island offers easy expansion.

YELLOW LIGHT

A fixed island should be far enough away from other cabinets and major appliances to allow for traffic flow and not hinder doors opening towards the island. Wheelchair access is also a consideration for some households. Recommendations vary, but figure a minimum of 36 inches of clearance outward from the island.

Freestanding Islands

- Cut out and tape a paper or cardboard template of your proposed island to the floor to get a feel for it as a work space.
- Include the bar stools or chairs in your measuring as they have to be walked around as well.
- A kitchen island's countertop does not have to match the rest of the kitchen—consider a contrasting material or one that's lower maintenance.
- Sturdy antique shop furnishings can be adapted as interesting movable islands.

Special Considerations

- When a sink is added to an island, it requires special venting to meet plumbing codes.
- For galley-type or other limited-space kitchens, countertop extensions are available for additional counter space.
- An extension typically folds against the cabinet at the end of the counter using a piano hinge, opens up, and locks in place when needed for food prep.
- Look for extensions at RV accessory suppliers.

ENERGY PRIMER

Learn daily habits to cut the cost of energy used in your home

When it comes to energy, we are well fed. We have bountiful electricity, heat, air conditioning, and hot water from reliable sources. Do we waste some? Of course, it's unavoidable; no system is perfectly efficient, but there are ways to conserve energy and cut down on your spending.

Oftentimes it's simply a matter of turning off a device we're not using—a light, the TV, a ventilation fan—instead of ignoring it. Other times it's a matter of degree: How warm or cool do we need the room temperature, how many lights do we need to perform a task, or do we really need an electric can opener instead of a hand-operated one?

Using and Losing Home Energy Sources

- Most home energy use goes towards heat.
- We heat water and the air, and all of our appliances create heat while they're running.
- Much of the electricity consumed by an incandescent light bulb becomes waste heat from the hot filament.
- Heat rises and is lost through roofs, both insulated and those lacking insulation; it's also lost through disconnected or poorly connected heat ducts, gaps around light fixtures, receptacles, and switches, as well as windows and doors with insufficient seals and weather stripping.

Typical Losses

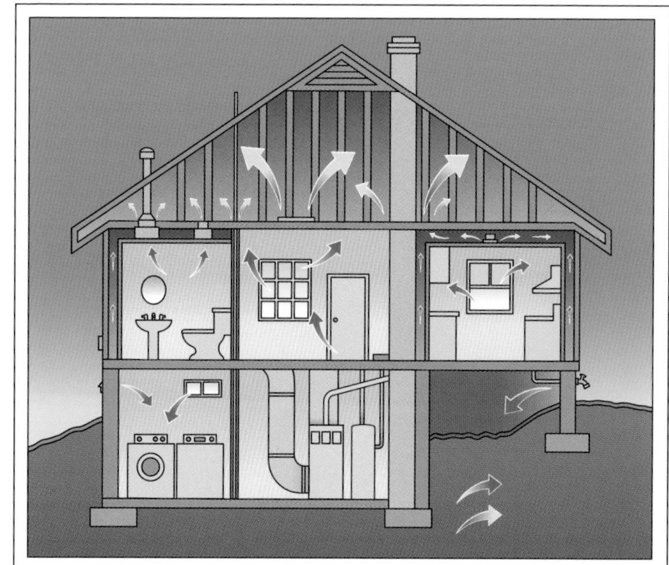

- According to Environmental Protection Agency studies, air leakage in the average American house accounts for 25–40 percent of the energy used for heating and cooling.
- Air leakage typically occurs at gaps around windows, doors, pipes, attics, joints, and chimneys without fireplace dampers.
- Sensors, dimmers, and timers can reduce lighting and individually running ventilation fan usage (and heat loss caused by ventilation).
- While controlling heat losses, a certain level of ventilation and air exchange is needed for a healthy indoor environment.

Adjusting to a slightly cooler home by lowering a thermostat one degree a month or insulating a water heater tank is not a lifestyle-bending change. Awareness is the key. Instead of continuing old habits, examine them and decide which need changing and which deserve to get booted out altogether. One approach is to limit how much you allow yourself to spend each month on energy and then find ways to meet your budget

MAKE IT EASY

Not sure where to start in evaluating your energy usage? There are professional energy auditors who, for a fee, will test your house for heat loss, check your furnace and air conditioning, and examine past energy bills. Your local utility company might conduct audits as well for free or for a small cost.

Water Heaters

- Water heating is the largest energy user in homes after heating and cooling.
- A conventional storage-style water heater tank should be sized to your hot water usage—too big of a tank wastes energy.
- Demand (tankless or instantaneous) water heaters provide hot water only as it is needed without the standby energy losses associated with storage water heaters; this can save you money.
- Demand systems cost more and last longer than conventional tanks.

Air Conditioning

- Reduce running your air conditioning and save energy and cost by using other cooling methods.
- Window shades, improved insulation, and sealing preserve cool interior temperatures just as they do warmer ones in the winter months.
- Individual, ceiling, and whole-house fans can increase summer comfort in hot climates.
- Evaporative (or swamp) coolers pull hot, dry outside air through moist pads, cooling the air by evaporation and using less energy than air conditioning.

SEALING YOUR HOUSE

Keep the heat and the cool where you want them—inside your home

We can't live in airtight homes. We need ventilation, but we don't need infiltration or air leakage. We can control ventilation with individual fans, both in kitchens and bathrooms, and whole-house fans. Control leakage with caulking, gaskets, and weather stripping. This is all low-tech, section-by-section tasking.

Individually, each missing seal doesn't seem like much, but cumulatively seals can add up to respectable savings. According to the U.S. Department of Energy, air infiltration can amount to as much as 30 percent of home heating and cooling costs.

Some claim that new homes can be too tight, that they don't "breathe" as in the good old days of less tightly con-

Sealing Switches and Receptacles

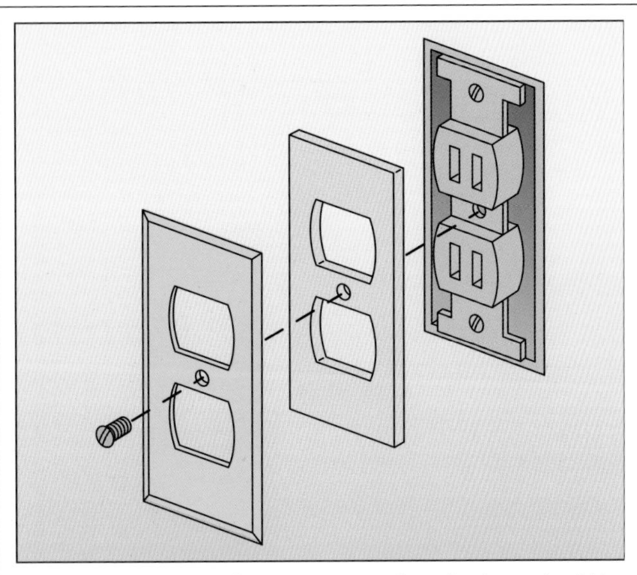

- Insulate switches and receptacles by removing the cover plates and installing foam gaskets, precut to fit (available at home centers).
- Seal individual room air conditioners with foam tape weather stripping installed between the air conditioner and the window frame.
- Use foam backer rod, sold in pieces and rolls of different thickness, to fill deep gaps between window trim and brick walls and then caulk right over the backer rod.
- Use a paintable elastomeric caulk for wood-to-masonry applications.

Attic Hatches

- Attic hatches should be both insulated and weather stripped around their perimeter.
- In addition to insulation, unfinished attics need adequate ventilation to prevent excess heat build-up and to allow moisture to escape.
- Self-powered solar attic fans do not require wiring to vent hot attics; this is an excellent use of solar energy.
- A properly ventilated attic helps reduce air conditioning costs and prevent roofing shingles from warping.

structed dwellings. New houses, however, exchange air in a controlled, timed fashion instead of leaking 24/7, like older houses often do through walls, windows, and roofs.

New house or old, most can use some inspection for air leakage. Even insulated attic spaces should be reconsidered as insulation standards have changed over the years. If the job seems overwhelming, break it down and do one task—weather stripping, for instance—at a time. Saving energy doesn't need to be complicated.

MAKE IT EASY

Older homes typically have open fireplaces, while newer ones have glass doors to prevent heat from going up the chimney. Keep the damper closed in an older chimney, while remembering to open it when you have a fire (a sign indicating "damper open" and "damper closed" is helpful) or install a tempered glass door.

Caulking

- Inspect the exterior of your house for separations between the foundation and the walls.
- Caulk these spaces with paintable, exterior aerosol insulating foam and paint the foam after it cures.
- Examine areas where corners or dissimilar materials meet, such as a concrete foundation and wood walls.
- Never caulk the drainage holes at the base of storm windows—these prevent moisture from building up between the storm sash and the main sashes.

Fireplace Doors

- If you have a fireplace with a pilot light, consider shutting it off for the spring and summer months.
- A gas pilot light costs up to a few dollars a month to run, depending on the appliance manufacturer.
- Relighting some fireplace pilot lights can be tricky—be sure you know how to do yours safely.
- Heat-resistant, tempered fireplace doors are both a safety feature and an energy saver and can be fitted to older fireplaces.

WEATHER STRIPPING

A range of weather stripping will seal up even the leakiest openings

Wherever you have an operable door or window, you have some air leakage. If you didn't, the opening would be so tight you couldn't open the door or window. Older sliding wood windows are the worst offenders. They were installed on-site, and no two installations were exactly the same. New windows are installed as finished, weather-stripped units, assembled according to uniform specifications in a factory, resulting in tighter tolerances and less air infiltration.

Old windows will almost always benefit from some type of weather stripping. Modern windows might need their weather stripping replaced, depending on its condition. Some homeowners caulk shut windows they're not using. This will

Weather Stripping

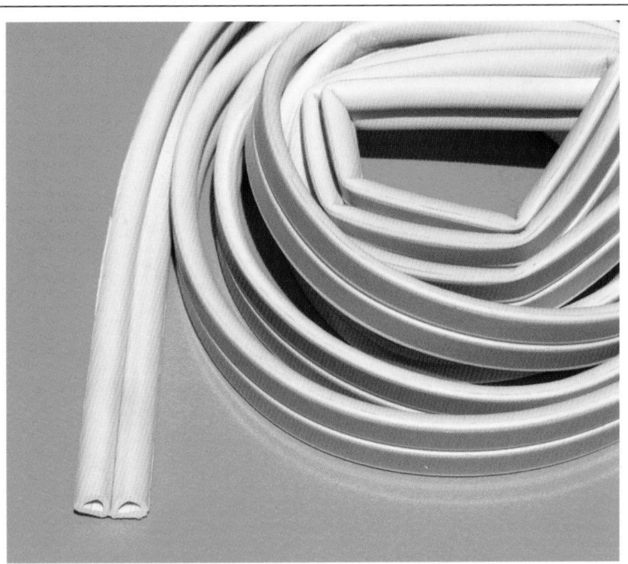

- There are multiple weather stripping options for every application, but some are much easier to paint around than others.
- An improper installation or the wrong weather stripping can prevent a door or window from completely closing and forming a tight seal.
- Avoid felt weather stripping—there are superior vinyl and rubber alternatives.
- New wood entry doors with self-adhesive vinyl weather stripping regularly have gaps; check yours for additional weather stripping needs.

Where to Install Weather Stripping

- Replace torn or missing weather stripping around entry doors with similar material or remove complete sections and install new.
- Check your weather stripping seals by turning off the inside lights at night and shining a flashlight around the doors and individual windows to see where light shines in.
- Garage door weather stripping kits contain special bottom seal weather stripping as well as jamb sealing weather stripping for the bottom and sides of a standard garage door.
- Some bottom seal weather stripping simply slides into an aluminum track, while other versions are nailed to the bottom of the garage door.
- Weather strip garage side doors as well.

certainly stop leakage but can be short-sighted. Do you really want to go without the summer ventilation each window offers because caulking is easier than weather stripping?

Newer homes have weather-stripped exterior doors but may not be completely sealed. Look for gaps and replace the stripping with something more effective. Weather stripping sliding patio or deck doors can be problematic depending on the type of door—wood, metal, or vinyl—and the original stripping. Replace worn or damaged stripping with material of similar dimension that won't interfere with the operation of the door.

ZOOM

For installation, the easiest weather stripping is self-adhesive foam, vinyl, or rubber. Metal weather stripping that requires fasteners for installation is the longest lasting and requires more precision when installing. Metal is also more forgiving when painting near it—paint spatters clean off more easily than off the self-adhesive products.

Metal, Rubber, and Vinyl Strips

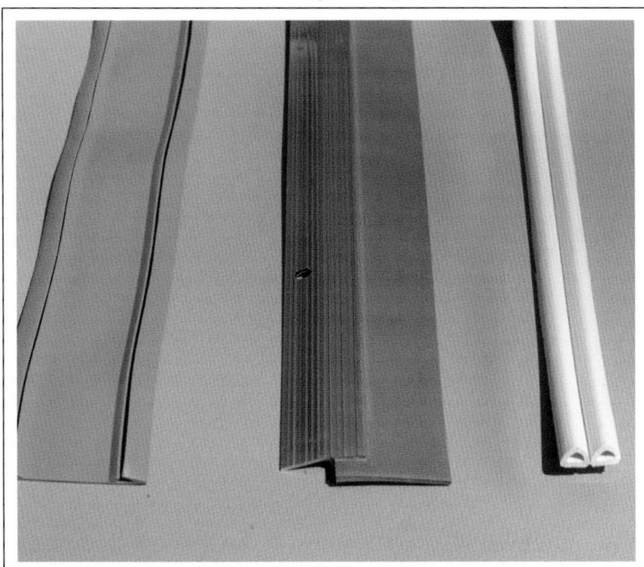

- Spring bronze metal weather stripping is extremely durable and a good option for old wood windows and doors.
- Self-adhesive vinyl weather stripping installs and sticks well to clean surfaces and can be easily replaced.
- Vinyl windows with damaged brush-type weather stripping can be repaired with self-adhesive vinyl weather stripping.
- Rigid metal nailing strips with rubber, vinyl, or silicone beads work when well-fastened to jambs and pressure fitted against closed doors.

Rope Caulk

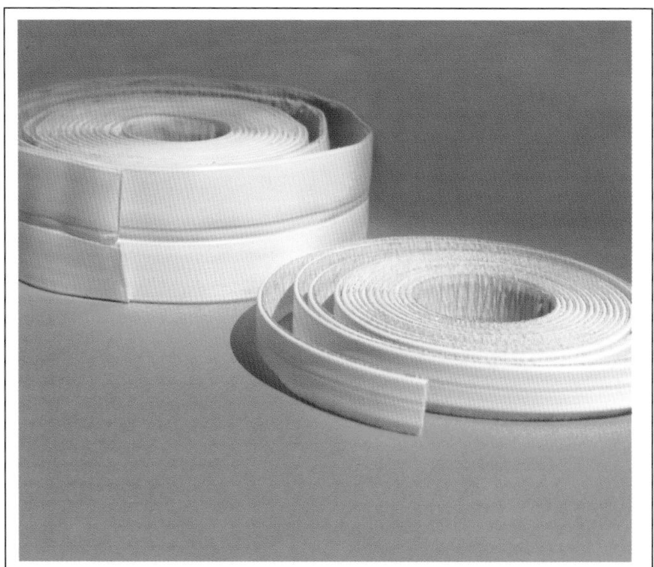

- Rope caulk is an inexpensive, soft, clay-like temporary caulking compound sold in multiple-strand rolls as an alternative to more permanent caulking.
- The strands are separated from the roll and pressed into gaps between windows and jambs in the late fall when windows are no longer opened and removed in the spring.
- Rope caulk stays flexible, adheres to clean surfaces, fills gaps, and requires no tools for installation.
- Once rope caulk is installed, the window should stay closed.

CHOOSING NEW WINDOWS

Depending on your need, storm windows might do the trick and help save energy

There's no escaping the advertisements for new insulated windows. Energy savings and low maintenance are promised, comfort is assured, and the cost is rarely mentioned. Vinyl windows dominate the new home and replacement markets, and there's much to be said for them: They never need painting, resist insects, and come fully assembled, and, unlike early white-only versions, vinyl windows are now available in a limited range of colors.

Some people are opposed to them because they're not wood and claim they have a limited life before the seals give out and leak, requiring the entire window be replaced.

Storm windows will effect the same or better energy sav-

The Ins and Outs of Windows

- Older windows, either metal or wood, have a single pane of glass, which doesn't meet modern building standards that call for insulated windows.
- An insulated window has two or three panes of glass sealed together with a special caulk, which also seals an inert gas between the glass panes.
- Heat loss occurs primarily through gaps between window sashes and their jambs.
- It takes years before the energy savings from new windows make up for the cost of the windows.

Wood Storm Windows

- Wood, aluminum, and vinyl storm windows are available, with wood being the most expensive.
- Some are fixed, meaning they do not open, while others have screens and movable glass or plastic sections for ventilation.
- Wood storm windows often push out at the bottom for modest ventilation.
- Vinyl interior storm windows offer ease of installation and cleaning but do not offer an exterior storm window's protection of the main window sashes.

ings as new insulated units, but there is the labor of installing, removing, and storing removable-style storms, and the less-than-appealing look of the fixed metal style. Interior storm windows are easier to maneuver than exteriors that call for ladders, but they do not protect the exterior side of the windows from winter weather, a big plus for exterior storm windows. New windows or storm windows—it's a practical and aesthetic decision.

GREEN LIGHT

Retaining original wood windows and adding storm windows not only maintains architectural integrity, but also reduces land fill refuse. Wood windows can last for decades with regular painting. The number one reason they deteriorate is a lack of paint. Keep them sealed, like all exterior wood, and they should never need replacement.

Aluminum Storm Windows

- Aluminum storm windows are lightweight and long-lasting, but as a metal they conduct heat instead of insulating against it.
- Wood sashes are heavier, require painting, and cannot accommodate built-in screens.
- Vinyl storm windows expand and move in hot, direct sunlight and can crack in extremely low temperatures.
- Unless they're easy to remove, storm windows will interfere with painting the windows, and heat build-up on wood sashes can cause the paint to blister.

Energy-Saving Costs

- Replacing windows or adding storm windows is an expensive way to save energy.
- Insulating the attic space and weather stripping older homes produce effective energy and cost savings.
- Addressing and tightening up the worst windows instead of wholesale replacement will also help.
- Selectively installing interior or exterior storm windows will increase the efficiency of those windows at a lower cost than full replacement while maintaining uniformity in appearance.

ADDING INSULATION

Check to see if your house is up to national insulation standards

According to studies done by Harvard University's School of Public Health, increasing residential insulation in existing housing to meet the 2000 International Energy Conservation Code (IEECC) minimum levels could save more than 800 trillion BTUs of energy each year. Translation: That means you will save more money. In addition, the environment will be cleaner because less fuel would be burned and thus lower emissions. Consult with a professional to make sure your insulation standards are up to par.

Blanket insulation is the most common and is normally made of fiberglass fibers attached to a kraft paper or foil facing. The facing acts as a vapor barrier that keeps the insulation dry.

Blanket insulation is sold in rolls, which you cut to length before installing, and as batts, which are precut 4- and 8-

When to Add Insulation

- It's worth doing a savings analysis to decide between keeping your present level of insulation and upgrading to a higher level.
- Adding insulation to the attic is easier than adding to the walls.
- Older homes might need more improvements because insulating was not a common or code-required practice until the 1970s, even in cold climates.
- When insulating a crawl space, be sure to cover any exposed ground with plastic sheathing to reduce the moisture.

Attic Blow-In Insulation

- An insulated attic prevents heat from escaping in winter and prevents outside heat from affecting air conditioning systems in summer.
- Pour in loose-fill fiberglass, mineral wool, or cellulose insulation or blow it in with a rented machine; the machine is easier and faster.
- Follow the manufacturer's recommendations for volume of material per square foot of attic space for proper coverage.
- Always wear protective respiratory gear when installing any kind of insulation.

foot sections. Both rolls and batts are wide enough to fit between floor joists and wall studs. The thicker the insulation, the higher its thermal resistance, or ability to keep heat from escaping. Insulation is also sold as loose fill or small bits of fiber, foam, and other materials that are blown into attics and wall spaces with commercial blowers. Insulation pays back immediately in comfort and energy savings.

RED LIGHT

When your blanket insulation faces a heated room—whether it's an attic floor facing bedrooms or an exterior wall facing living space—the facing or vapor barrier should face the heated area. If you're adding blanket insulation on top of existing attic insulation, remove the facing or use insulation without facing.

Wall Blow-In Insulation

- You can blow insulation into the walls of older homes without removing either the siding or the interior plaster.
- Drill a hole near the top of the wall and blow the insulation between the wall studs.
- Before blowing insulation into the outer walls, complete any wiring and plumbing projects, as you can't pass wires or pipes through the material once it's installed.
- Avoid using blown-in insulation near chimneys or wall sconces.

How to Blow In Insulation

- Keep the machine outside to avoid blowing insulation dust inside your house.
- This is a two-person job—one to feed material to and control the machine and one to hold the hose to the holes and fill the walls.
- Wrap a rag around the end of the hose to prevent insulation from blowing back into the room.
- Fill the interior holes with aerosol foam filler, cut the dried filler flush to the wall, and smooth over with drywall compound.

APPLIANCES & ENERGY SAVINGS

Look for energy-saver appliances to help lower your monthly bills

Appliances definitely make life easier, but you don't need them to suck out all your house energy. New energy standards—especially the ENERGY STAR program, a set of energy efficiency guidelines set by the EPA and U.S. Department of Energy—have improved some of the major power consumers.

Kitchen ranges offer more oven insulation than previous models and automatic pilot lights for gas ranges. New ENERGY STAR-rated refrigerators use less than one-fourth as much electricity as they did in 1972. A modern ENERGY STAR clothes washer uses less water and electricity than models manufactured a mere fifteen years ago, and ENERGY STAR clothes dryer models use less electricity. Should you replace

Install a Programmable Thermostat

- Programmable thermostats are standard in new homes and can be retrofitted to many older systems.
- Call an electrician to install any thermostat running off high-voltage wiring.
- Turn off the power to the thermostat, furnace, and air conditioner before removing the old thermostat.
- Mark the old wires according to where they were attached on the old thermostat and tape them to the wall to prevent them from falling into the wall cavity.

Water Heater Temperature

- According to the U.S. Department of Energy, for each 10-degree reduction in water temperature, you save 3–5 percent in energy costs.
- A hot water temperature of 120 degrees F is recommended for most households, although some dishwashers might require 130 degrees F or more.
- If you have an electric water heater, consult the owner's manual for precautions for adjusting the thermostat.
- Insulating accessible hot water pipes reduces heat loss and can raise water temperature 2–4 degrees more than noninsulated pipes.

for the sake of efficiency? It depends on the age of the appliance you're replacing and the amount of energy it consumes versus the cost of a new appliance and the cost savings from its lower energy use.

You can further cut down electricity consumption and costs by using your appliances smartly. For instance, run full loads of clothes and dishes or use a refrigerator to its full capacity.

GREEN LIGHT

Do you really need to keep your old refrigerator in the garage or basement as an "extra"? This more than negates any energy savings with a new model. Better to buy a larger kitchen refrigerator if you need the extra capacity than maintain a second one.

Lowest-Flow Showerheads

- Showering accounts for 25 percent of individual water usage.
- Low-flow showerheads with a flow rate of 2.5 gallons per minute or less include aerating, which mixes air with water, and nonaerating models.
- Another option is a showerhead with a built-in shut-off or soap-up valve, which allows the user to decrease the water flow down to a light mist or shut off completely.
- Installation is easy and requires only a wrench and Teflon plumber's tape.

Exhaust Fans

- Energy saving and exhaust fans are somewhat mutually exclusive.
- You want a powerful fan that draws fumes and moisture out fast, which means it removes heat quickly, too; it's better to vent faster and for a shorter amount of time.
- Run your exhaust fans only as long as you need to clear the bathroom and keep heat loss to a minimum.
- A bathroom fan timer can control and reduce excess usage.

FENCE REPAIRS

Wood fences can't stand up forever—they need help from time to time

Wood fences go back to the early colonists. Picket fences were built to establish boundaries but not privacy. Some were quite elaborate, much more so than most anything you'll find today.

Solid fences built for privacy are popular now and simple to construct: 4″X4″ posts are sunk into the ground no more than 8 feet apart, 2″X4″ boards are nailed to the posts as rails, and then 1″X6″ boards, frequently rough cut, are nailed to the rails. More often than not, these fences are then stained or even left alone to "weather."

This type of fence doesn't require elaborate carpentry skills to build, but once it's up, it requires care and feeding. Wood

Signs of a Failed Fence

- A fence falls into disrepair when the posts rot, the post holes aren't deep enough, the rails come loose from the posts, or the individual fence boards come loose.
- Unsealed fences deteriorate faster, and they often have loose boards and fasteners.
- Wood fences are expensive and time consuming to build—it makes good sense to maintain them.
- Figure on restaining a fence every three to four years depending on weather exposure.

Loose Fence Boards

- Loose boards can be secured with small deck screws or galvanized nails.
- If an individual board is broken or otherwise too deteriorated to secure, new ones are available at any lumber store.
- Prestain or paint any replacement boards before installing.
- Be sure the ends of all the fence boards are well sealed and none comes in contact with the ground.

fences are exposed to the weather and need to be treated like any other outdoor wood. They must be sealed regularly with paint or stain and checked for deterioration.

Traditional picket fences are painted, providing them with better protection, if the paint is regularly renewed, than stained solid fences, which are often ignored. Given the cost and time involved to construct a fence, recoating to extend the life of a fence more than pays for itself.

MAKE IT EASY

Recoating a fence is time consuming, but the task can be sped up using a paint sprayer, particularly when painting a picket fence. An airless sprayer will paint pickets far faster than brushing, and overspray is less an issue outside in a yard than inside your house. Solid fences can be rolled or sprayed.

Reinforcing Loose Rails

- Rails that are nailed at an angle (toenailed) to the posts commonly come loose.
- Reinforce a loose rail by nailing a small wood block to the post; butt it up against the rail and toenail the rail to the block.
- Or insert a galvanized angle iron under the rail and screw it to both the rail and the post.
- A nailed section of 2X4 across the post and into each rail produces the sturdiest but most visible repair.

Fence Caps

- Fence caps become loose when they remain unfinished and the nails rust and deteriorate.
- Remove loose post caps and their fasteners. Sand the cap's bottom side and the top of the post. Brush away dust and seal all sides of the cap with wood sealer.
- When the cap is dry, apply exterior wood glue to the top of the post, and center the cap on it.
- Secure with two deck screws or two galvanized nails longer than the original ones.

FENCE REPAIRS (CONTINUED)

Fences in disrepair need more than recoating—posts, rails, boards, and pickets can all be fixed

A properly built fence uses posts that are treated for ground contact. Lumber manufacturers pressure-treat lumber with chemicals that render the wood resistant to fungi found in the ground. Some pressure-treated lumber is not suitable for ground contact and should not be used for fence posts, nor should untreated lumber be used. In the event of rotting wood, individual posts can be replaced or supported and braced until replacement can be scheduled.

Rails eventually loosen if they don't have any supports to hold them or if the nails securing the rails corrode. The same is true for the pickets or vertical boards. A section of a fence either can be secured with additional fasteners or replaced

Bad Fence Posts

- Reinforce fence posts with inexpensive fence-post repair bracket kits, which save the time and cost of post replacement.
- Repair kits require existing concrete (to be hammered into) or new concrete if the post stands in dirt only; they do not work for posts with extensive rot.
- To quickly repair a leaning post, screw one end of a wire or small chain near the top of the post. Straighten the post, pull the wire tight, and attach to a pipe hammered into the ground.

Removing Rails and Posts

- Rails are either toenailed (nailed at an angle) to posts or nailed to metal clips, which are nailed to the posts.
- Remove the rails and temporarily nail a vertical 2X4 to each pair of rails to keep the fencing upright.
- Dig out the old post and any concrete attached to it.
- If a rail end is bad, cut out the bad section, install a new piece of 2X4, and secure it to the remaining 2X4 with a galvanized shelf bracket.

completely if necessary. This can be a small problem if a picket is an unusual design that requires duplication. However, new 1″X6″ fence boards are commonly available at any lumber store.

Anywhere a fence is feeling wobbly or is otherwise in need of support, brackets, blocks, cables, or other means of shoring up the weak area can be installed to extend the life of the fence. As long as the repair is presentable, it's simpler than a full replacement.

YELLOW LIGHT

Treated lumber is soaked in a chemical solution that will rub off on your skin during handling. Wear work gloves and long-sleeve shirts when working with treated fence posts. Wear an appropriate dust mask to avoid breathing any sawdust while cutting the posts. Wash your hands after handling treated lumber.

Digging Post Holes

- Dig post holes with a narrow shovel, a post-hole digger, a hand-operated auger, or a handheld power auger (but beware of usage hazards).
- An alternative is a 24- or 30-inch fence post spike topped with a 4″X4″ open metal box for securing a fence post.
- Before digging holes for a new fence, call your local utilities to check for pipes and wiring (the North American One Call Referral service provides a listing of local utility companies; see Resources).

Positioning and Installing Posts

- A post hole should be at least 24 inches deep for a 6-foot-high fence.
- The hole should also be wide enough for you to pack down your dirt or concrete packing.
- The new post should be straight, lined up with its fence rails and with the other posts; if the posts are on sloped ground, it should follow the slope as the others do.
- Allow concrete to set before attaching the rails, following the manufacturer's curing directions.

DECK REPAIRS

With some attention, decks can last for years—here's how

A lot is asked of a deck. It's laid out horizontally so water can't run off and is lucky if it gets recoated every few years. It bakes in the sun and freezes in the winter. Luckily, like any other wood structure, each damaged component of a deck is replaceable.

A deck that's seen better days but is structurally solid can always be cleaned and refinished. What if it's rough and splintered? Sanding a deck will bring it back to life and prepare it for a new finish. Deck boards that are too far gone to be saved can be removed and replaced easily enough. Not as easily replaced are deck joists, the supports that the deck boards are nailed into. Small areas of each joist edge are exposed to the weather and usually don't directly receive any finish during

Replacing Deck Sections

- Decking and deck rails are frequent candidates for replacement due to water exposure.
- Each piece of decking, typically a 2X4 of some kind, is nailed twice at each joist.
- With a large pry bar, remove any damaged boards, being sure to place a block of wood under the pry bar to avoid marring the adjoining boards.
- Replace with the same dimension and wood type and stain to match the rest of the deck (see page 38).

Repairing or Replacing Joists

- If a section of a joist is deteriorated, cut away the damaged area with a reciprocating saw and coat the exposed wood with a preservative.
- Cut a new joist the same length as the damaged joist.
- Glue and fasten the full length joist to the old and secure at the ends in the same manner as the others.
- If joist ends are damaged, cut away the damaged material, coat the exposed wood with preservative, and run a new joist the entire length of the old.

recoats as they're not easily accessible. Not surprisingly, these areas can deteriorate.

Posts are also main supports but are replaceable or can have new sections spliced to them. Replacing an entire deck is expensive and a lot of work, so you might consider hiring a deck contractor. Refinishing and patching the neglected areas will go a long way in extending the life of a deck.

YELLOW LIGHT

Before sanding a deck, set all the nails in the decking. Often times the nail heads are sticking up far enough to catch a piece of sandpaper and tear it. You'll have to set the protruding nails anyway after tearing your sandpaper, so save yourself the aggravation and set them first.

Firming Up Decks

- Decks get wobbly due to age or insufficient bracing, especially second-story and higher decks.
- There is no limit to the amount of bracing and support you can add to a deck; more rather than less is a good idea.
- Use deck screws to secure 2X4 braces to deck posts or joists to stiffen up the structure; nail or screw sections of 2X6s between joists as well.
- Some decks are built with undersized or minimal number of posts—install additional ones for stability.

Repairing or Replacing Deck Railings

- Deck railings should be functional first and decorative second while meeting local safety codes.
- The top railing is subject to deterioration and can be replaced with a new 2X4 or 2X6 as needed (attach with deck screws).
- Prefinish any replaced parts with stain or paint to match the rest of the deck.
- Replacing an entire railing and supports will update your deck and can be done with a different material, including low-maintenance PVC or metal.

OUTDOOR LIGHTING

With a little lighting, your yard becomes a lot more usable

In the summer months, our yards become our personal playgrounds and weekend gathering spots during the daylight hours. When night rolls around, a little extra lighting allows for extra use of the back yard. A yard can be as lit up as a kitchen or as subdued as a romantic restaurant, depending on what you like.

Extra lighting can also bring a sense of security, particularly for those living alone and concerned about entry doors and entry-level windows. The types of lighting vary greatly.

Outdoor lights can be mounted on a structure, on or at the base of trees, or at any ground location. Hard-wired lights require trenching in order to run the wires but offer a great

Hard-Wired Lights

- Hard-wired outdoor lights offer the brightest illumination options but are the most difficult to install.
- Wiring must meet local electrical codes, and the work must be inspected and approved.
- If an interior electrical circuit can accommodate an additional load (see page 112), an exterior mounted light can be wired to it instead of running new wiring from the service panel.
- In addition to outdoor lights, consider adding outdoor-rated electrical receptacles for convenience.

Wireless Solar Lights

- Solar outdoor lights are the easiest to install and require no electrical permits.
- Full lines of lights and styles are available, including path lights and spotlights.
- Other than occasionally wiping the solar panel clean and replacing the batteries every one to three years, there is no maintenance for solar outdoor lights.
- Because they're not permanently wired, these lights can be relocated to other parts of your yard.

variety of lighting styles and brightness. A wireless solar light has a small solar panel. This panel absorbs daylight, which charges an internal nicad battery. The battery powers an LED, which provides a small amount of illumination, nowhere near as much as a wired light can provide.

Which type of lighting is best? It depends on what you're trying to do. Solar lights do a good job marking sidewalks and driveways, and wired lights let you see more, an important consideration for security. Combining both offers more options with less compromise.

YELLOW LIGHT

Underground wiring can be installed in metal or plastic conduit or a direct burial cable. Each requires a different burying depth. The location of your lights should be planned with the wiring path in mind. Some paths will be more difficult to dig than others or be more susceptible to future disruptions (tree roots, for instance).

Motion Detectors and Timers

- Motion detector-activated lights allow lighting when you need it instead of remaining on all night.
- For security lights, install them high enough that an intruder can't disable the motion detector mechanism or remove the bulbs from the fixture.
- Timers allow you to control designated lighting, both indoor and outdoor, according to seasonally adjusted darkness.
- You can control holiday displays or fountain pumps with timers, too.

Location of Lighting

- Install lights for safety along high-traffic areas and especially near steps, which are hard to view in the dark.
- The location of your power source will affect where you install hard-wired lights.
- The type of lighting you want in any specific location will determine the design and style of the fixture.
- For pure decoration, choose a focal point or two, such as a large tree, for additional lighting.

COMPOSTING

Why pay to have yard debris hauled away when you can easily reuse it?

Nature recycles itself. Plants die and decompose, enriching the soil. Seeds blow or drop in, and the cycle starts all over again. In an effort to better the environment, cities have increasingly called for separating yard waste from household trash for large-scale composting. Homeowners can do the same thing and save the hauling fees.

What can you compost? Most any organic material, but they're not all created equal. Leaves, grass clippings, wood (but not coal) ashes, kitchen vegetable and fruit refuse, manure from vegetarian animals, seaweed, and even dryer lint are all good compost candidates. The smaller the material can be ground or chopped up, the faster it will decompose.

What Is Composting?

- Composting and the general breakdown of organic material work best with a balanced carbon (dried leaves, wood chips, straw) to nitrogen (grass clippings, kitchen scraps) ratio among the compost pile's ingredients.
- Too little nitrogen slows decomposition while too much speeds it up and creates an odor problem as oxygen is consumed too quickly.
- Aim for about 25 parts carbon to 1 part nitrogen, but don't worry about a perfect mix.
- Regularly water and mix for aeration.

Different Composting Bins

- A compost bin can be an unsecured heap, a sealed plastic bucket, or a fenced-in corner of your yard.
- The incorporation method calls for small amounts of nonfatty food wastes (no meats or bones) to be mixed with soil and directly buried 8 inches down or deeper.
- Plastic compost bins work well to control vermin attracted to compost; they are available in tumbler models for churning the contents.
- An open wire-mesh compost bin is appropriate for yard waste, less so for food scraps.

A compost pile requires as much or as little work as you wish. A pile left to its own will eventually break down, while a managed pile that you rotate and water will decompose much faster. Concerned about unsightly piles of dead stuff in your yard attracting vermin? Closed plastic compost bins are available in several styles and sizes. Set one up in a corner of your yard, save the city some truck fuel, and enrich your garden at the same time.

MAKE IT EASY

Grass clippings can be recycled by leaving them on the lawn. Clippings add nutrients, help with water retention in hot weather, and reduce the need for fertilizer. Short clippings work best, which can mean more frequent mowing or the use of a mulching lawn mower. Clippings can also be used in gardens.

Preventing Rodents and Pests

- Keeping your compost wet and turned helps prevent rats from nesting in it.
- Limit food additions to the compost and bury them deep so they're harder for rats and other vermin to dig out.
- Consider two compost bins, one open pile for yard waste and a second enclosed bin for food scraps.
- If necessary, enclose an open bin with 1/4-inch wire mesh, buried down several inches to prevent rats from burrowing into the compost.

Worm Bins

- Worm composting can reduce food waste and create rich compost.
- In an aerated plastic container, your worms can live indoors or out as long as the temperature is 40–80 degrees and they're kept out of direct sunlight.
- Each bin requires moistened bedding, redworms, and a supply of nonfatty food scraps, especially fruit, vegetables, crushed eggshells, and coffee grounds.
- Minimal maintenance includes refreshing the bedding and removing the compost a few times a year.

RAIN BARRELS

Save rainwater for a sunny day and help the environment

According to the American Water Works Association, the daily indoor per capita water use in the typical single-family home is 69.3 gallons. Billions of gallons of water could be saved every day if water-efficient fixtures were installed. Rain barrels aren't considered water-efficient fixtures, but they indeed save water by cutting down on the demand for piped-in water. One inch of rain falling on 1,000 square feet of roofing is up to 500 gallons of water.

Some rain-collection systems are nothing more than a plastic barrel into which a downspout is inserted. Water collected between periods of rain can supplement tap water for garden and lawn use. Other systems are quite sophisticated.

Why Use Rain Barrels?

- Rainwater is free and does not go through any processing; it's often soft water without dissolved minerals.
- Rain barrels can be located in convenient areas for watering your garden if a spigot isn't nearby.
- Extensive use of rain barrels would keep unnecessary water out of treatment plants and decrease sewer flow during heavy storms.
- Multiple barrels can be combined for increased capacity, especially during periods between rainfalls.

Rain Barrels of Any Style

- Rain barrels are available for purchase in approximately 50–80-gallon sizes and a variety of styles and designs.
- You can make your own using a 55-gallon food-quality recycled barrel, which will require a spigot and an overflow control be installed.
- Rain barrel water is not potable but can be used for watering indoor and outdoor plants and washing outdoor furniture, wheel barrows, etc.
- Keep barrels covered at all times to prevent mosquitoes from laying eggs in the water.

They include large storage tanks and filtration systems rendering rainwater reusable for washing machines and toilet flushing and in some cases as drinking water.

Installing and maintaining rain barrels are not complicated. Various vendors offer barrels in a range of sizes (typically 50 gallons to over 100) with spigots and overflow valves. Connect a hose to each, and excess water is directed away from the barrel and the house foundation, and the remaining water stands ready for later use. It's uncomplicated while reducing demand on your local water facility.

ZOOM

Colorado law disallows rain barrels due to water rights laws. Unless you have a right to the water falling on your property you can't keep it. Water in Colorado is assigned to water right holders, and rainwater prevented from running into streams and rivers is not going to its rightful owner.

How Much Water Can You Collect?

- According to the Southwest Florida Water Management District, figure about a half gallon of water per square foot of roof area during a 1-inch rainfall.
- A 2,000-square-foot roof can collect about 1,000 gallons of water.
- Multiply the outside length of your house by its width to get a rough idea of your roof's catchment area.
- Check the online figures for your area's typical rainfall to calculate potential water collection volume.

Saving Money and Water

- Rain barrels will not provide enough water or water pressure to water a large lawn but can water gardens and supply soaker hoses.
- Some municipalities offer rain barrel rebate programs as well as information on where to purchase rain barrels and how to install them.
- Rain barrels are not always available at garden and home supply stores—call ahead first.
- Kits for converting or retrofitting a barrel might be stocked at these stores more often than the barrels.

SAFETY GATES

Set one of these up to keep young children out of harm's way

Young kids like to wander—and they should. Everything is interesting to them. Once they become mobile, little stops them from snooping around. You want to keep them safe, of course, without hindering them too much. Safety gates restrict access but not curiosity, and they'll keep kids from going into unsupervised parts of the home. They aren't substitutes for parental care and oversight, but they're a terrific parent helper.

That said, many safety gates aren't all that attractive. Different models come with different means of attaching them to walls and woodwork (some of the pressure type rubber contacts can pull paint off woodwork unless you put thin

When to Install a Gate

- As soon as your child gets active, it's time to get safety gates.
- You want the gates available and ready to install once crawling begins, not finding yourself scrambling around trying to block off doorways without them.
- Gates aren't for children only—use them when you train a new puppy or even an older dog.
- Gates are not a substitute for child-proofing your house but a helpful addition.

Safety Gates

- Pressure-style safety gates are appropriate for doorways with solid-wood jambs to press the rubber ends against for a tight fit.
- These do not require any tools for installation and are expandable to fit different size openings.
- If the gate is moved daily, consider a hardware-mounted gate instead.

cardboard against the paint first). Today online sellers of baby safety products offer gates for every configuration and need, including furniture quality hardwood models. Safety gates also keep pets where you want them and prevent elderly persons with mental impairment from leaving safe areas of a home and possibly leaving altogether.

ZOOM

Safety gates come in two categories: pressure-mounted and those with screw-in hinges for attaching to woodwork and door frames. Be sure your gate works for the intended location. Pressure-mounted gates, for example, are inappropriate for the tops of staircases as they can be pushed over. For nonstandard-size openings, adjustable gates are available.

Hardware-Mounted Safety Gates

- A hardware-mounted gate is secured with screws and hardware to a wall stud, post, or door jamb.
- At the top and bottom of staircases, hardware-mounted are the only safety gates considered to be secure.
- Installations at the top of a staircase should keep the gate from swinging towards the steps.
- To avoid installing hardware in a finished wood post, consider tightly clamping a 2X4 board to the post and fastening the gate to it.

Where to Install

- Install safety gates at the top and bottom of any open staircase.
- Install the gate where it has sturdy surfaces to mount to or against.
- If a wall stud isn't available or you want to minimize damaging wallpaper, install a very tight floor-to-ceiling 2X4 board with a long enough 2X4 nailed to the top to brace against one or two joists in the ceiling; use a scrap carpet at the bottom end.

WINDOW PRECAUTIONS

Make windows in your home child-safe while keeping them functional

For the most part, kids are careful around windows, but accidents happen. According to the Consumer Product Safety Commission, more than four thousand children end up in emergency rooms every year due to window-related falls.

The basics should be addressed first: Move furniture that kids might play on away from operable windows so kids don't accidentally tumble out and be sure they generally play away from windows, even when they're closed; don't depend on window screens to prevent falls; if you're installing new windows, consider how you want them to open (top openers are safer).

What about old windows? Double-hung windows open from the top and bottom. Freeing up the top sashes provides plenty

Limiting Window Openings

- In warm weather, be aware of open windows, which could prove dangerous for small children, particularly in multiple-story apartment buildings and condominiums.
- In buildings without air conditioning, limit window openings around children to 4 inches for safety and adequate ventilation.
- Windows without limiting devices should be open only when children in those rooms are monitored.

Types of Guards

- Hardware is available that limits how far a window sash will open, while other styles are movable, allowing optional full opening.
- Casement windows can be controlled with a standard door chain lock, installed high enough so a child cannot reach it.
- Interior window guards offer the fullest protection and ventilation.
- Exterior guards are intended as antiburglary devices and are fixed in place unless they remotely unlock for escaping fires.

of ventilation. Installing a hardware stop can keep the lower sash closed until you want to open it, or a safety guard can be installed. Casement windows also accommodate safety guards.

Pivoting windows cannot be adapted to conventional window guards, but their movement can be limited with blocks, a door chain lock, or other hardware attached to the jamb. Pivoting windows are uncommon but are occasionally found in some older homes. Take the right precautions to keep your children safe.

RED LIGHT

You must be able to remove child protective guards and locks quickly should your family need to escape a fire. Older children and all adults should practice removing and releasing them. We have far fewer fires than twenty years ago, but never take a chance that one won't occur in your home.

Vent Locks

- Safety locks work on sliding windows, including double-hung style, as well as sliding patio doors.
- Some ventilation locks have a movable pin or other metal section that prevents a sash from opening until the pin is moved.
- These locks are inexpensive and easy to install.
- Some styles of vent wedges adhere using their own adhesive and are meant to stay fixed in one location.

Window Blind Cords

- Window blinds whose pull cords have loops have been identified as a strangulation hazard to young children.
- Pull cords with loops were eliminated, but older products can still have this feature.
- As a free service, the Window Covering Safety Council provides consumers with free cord stops, tassels, tie-down devices, and a safety brochure with retrofit instructions to prevent child-related accidents (see Resources).
- As an alternative, install cordless window coverings in children's rooms and playrooms.

ROOM & CABINET DOORS

With specialized hardware, doors can remain secured while your kids stay safe

Cabinet door latches have been around for years. In addition to keeping prying eyes and hands out—and a cabinet's contents in—safety latches are useful in the event of an earthquake in preventing these same contents from spilling out. Very young children like to empty cabinets and drawers. Latches keep things in order as a convenience to you, although some parents choose to keep one kitchen cabinet available for exploration.

Locks and latches come in different styles, most of them hidden behind cabinet doors or drawers, while others lock straight through side-by-side cabinet door handles.

Locks and latches also prevent room and entry doors from accidentally being slammed on little fingers or toes. Finger

Protecting Fingers

- Doorknob guards and locks prevent small hands from opening doors, but consider any elderly household members who might have trouble with these devices.
- Latches and locks installed above the doorknob are out of children's reach yet maneuverable by adults.
- Plastic door stops and plastic slip-on devices prevent doors from slamming into walls when opened and from damaging fingers near the door jamb when closed.
- Door-positioning hardware installed at the bottom of the door will keep it open and prevent it from closing accidentally.

Cabinet Locks

- Plastic latches that mount on the inside of cabinet doors and drawers keep both accessible to adults but not children.
- Some drawer latches require only self-adhesive installations, and others allow the drawer to open slightly and then lock in place to avoid pinched fingers.
- Surface-mounted U-shaped plastic locks secure side-by-side drawer knobs or cabinet knobs, preventing their opening.
- Self-adhesive oven and refrigerator door locks are also available.

guards of various types are available to prevent doors from completely shutting and slamming onto vulnerable fingers.

For bifold doors, a very simple and inexpensive plastic stop inserted at the top of the doors where they come together prevents them from opening. Locks are also available for sliding doors. If a door needs to stay open, door-positioning hardware installs on the bottom edge to prevent movement.

Just a few simple adjustments to your doors will help prevent accidents.

RED LIGHT

Installing safety latches doesn't guarantee a child won't get past them and grab at toxic cleaners and chemicals kept in lower cabinets. Dangerous products should be kept out of reach of children—period. When your children are old enough to understand the dangers of these products, you can move them.

Locks

- Child safety locks are inexpensive and easy to install.
- The locks must be deterrents to children as well as workable by adults—this might not be the case if arthritis prevents easily working the lock.
- Besides storing toxic cleaning substances in higher cabinets, consider storing breakable glass and sharp knives and scissors away from curious children, too.
- Secure medicine and medicine cabinets as well.

Bathroom Emergency Door Keys

- Newer bathroom door locksets secure with a button or lever to activate the locking function.
- The outside of these locksets have a small hole in the doorknob for inserting an emergency or privacy key, which should come with the lockset.
- These inexpensive keys are sold separately where locksets are sold—be sure to put the key in a known place and use if your child accidentally locks the bathroom door and cannot open it.

ELECTRICAL PRECAUTIONS

Take preventative measures against electrical mishaps to ensure the safety of your family

Young children love poking fingers, pens, and keys into electrical receptacles. It's easy and inviting, given receptacles are so close to floor level. Over a thousand children, most four years old and younger, get treated each year in emergency rooms for injuries from playing with receptacles, according to the U.S. Consumer Product Safety Commission.

The oldest preventative measure is inserting plastic plugs into unused receptacles, but newer products are available to maintain safety and allow easier use of receptacles. Some completely cover receptacles and the cords plugged into them. Others protect fingers from a plug if it's partially pulled out from a receptacle. For a modest cost, you can retrofit any or all of your

Cord Organizers

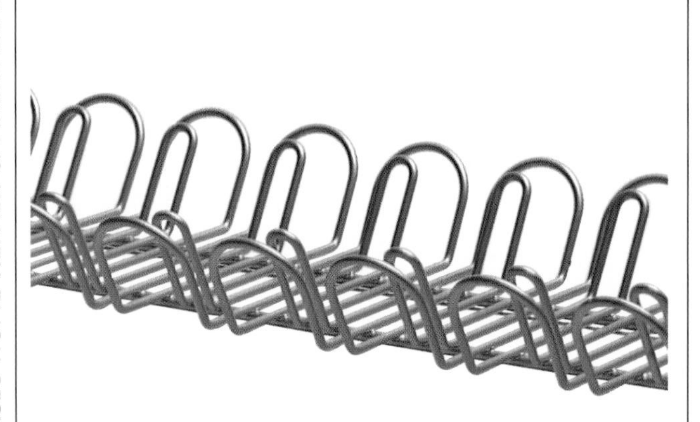

- Organizing multiple electrical cords behind computers and electrical systems benefits children and adults by keeping wires neat and safe.
- One approach is a plastic tube-type product that encases multiple cords in a neat grouping and away from small hands.
- Another product encases cords in a hard plastic container and winds them up so a cord is less of a tripping hazard.
- Self-adhesive plastic cord covers enclose single wires to both protect and hide them.

Securing Loose Wires

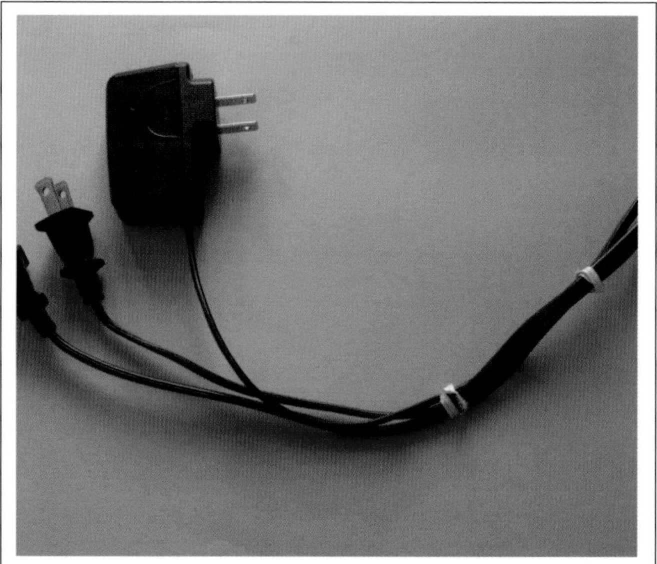

- Although not as protective as using cord covers or casings, taping or binding multiple cords together with wire twist ties will make individual cords difficult to pull at or chew on.
- Use adhesive-backed wire clips or screw-in cable clamps to secure wires to baseboards.
- Replace any frayed or worn cord for the family's safety.
- Don't leave plugged-in kitchen appliances unattended when young children are present.

receptacles and protect your children from shock and burns.

Cords are another attraction, especially when babies are crawling. Cord organizers are available for multiple cords (sound systems, computers and peripherals, etc.) as well as protective coverings for cords running along baseboards.

One of the safest actions you can take is to have your electrician install switch-controlled receptacles. Your receptacles are controlled by a wall-mounted switch the same way a ceiling light is controlled. Unless you're in the room to monitor your children, the power can remain off.

YELLOW LIGHT

No amount of protective devices can replace an attentive parent or guardian. Like cabinet latches and safety gates, these devices are a big help but only supplement a watchful adult. Secure your electrical area as best you can but understand the limitations of the hardware and devices you install.

Power Strip Safety Cover

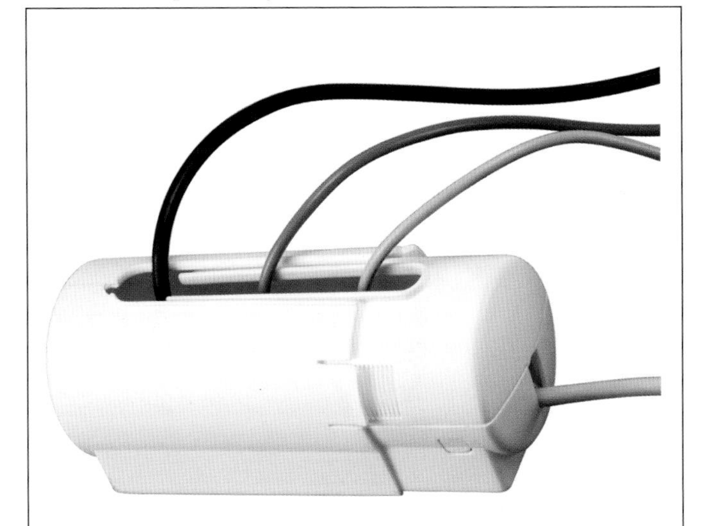

- Power strips that feature multiple outlets have safety covers available, allowing all cords to be safely plugged in and inaccessible to small hands.
- These covers also work for surge protectors.
- These are not only a safety device but also prevent curious children from yanking all the plugs out of a strip.
- Check that the cover you're buying will fit your power strip—there is some slight difference in size among the available covers.

Outlet Covers

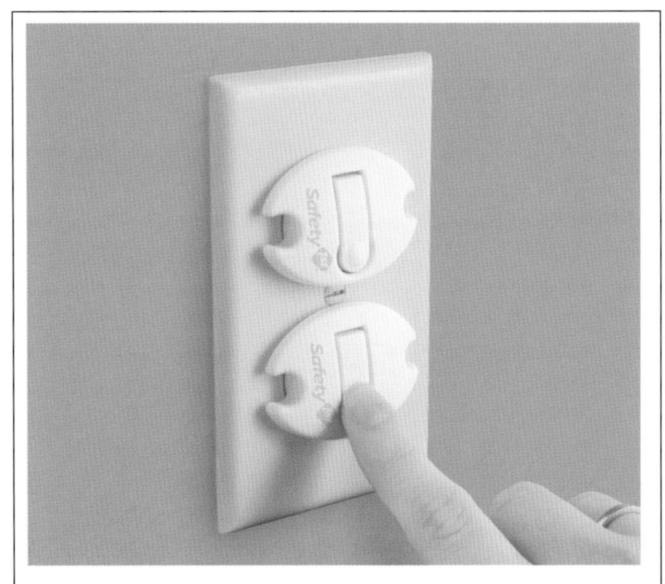

- The most basic outlet covers are traditional plastic plugs that insert into a receptacle; they're effective and inexpensive but must be removed before use.
- Some covers allow one or two cords to be plugged into the receptacle while preventing them from being touched.
- Another style uses a sliding mechanism to cover the receptacle—it moves aside when a plug is inserted.
- One high-tech cover accepts plugs, detecting them with sensors before allowing electricity to flow to them.

PLUMBING SAFEGUARDS

Control the flow of water to stop messes and potential burns on children

Kids love water. It's fun to splash and interesting when it keeps coming out of a tap. Toilets are interesting, too. Push a handle, and things disappear. Toys and washcloths also disappear. As much as you want to limit water play to bath time or outside with a hose, your kids have different ideas, but you can control the water flow and extracurricular toilet usage with some readily available kid-proofing hardware and devices.

A key to child safety around water inside your home is checking the water temperature. According to the McKesson Pediatric Advisor, it takes only six seconds of exposure to 140-degree F water to cause a serious burn but ten minutes

Toilet Lid Locks

- Self-clamping toilet lid locks prevent the youngest children from opening the lid but not adults or older children.
- For the price of one visit by a plumber, you could buy a dozen locks or more, so they are easily worth the investment.
- Different styles are available, but some do not fit all toilets as their promotional material suggests.
- No tools are required to install toilet lid locks.

Safety Tap

- Even with water heater tanks set to lower temperatures, children can still get injured by uncontrolled hot water coming out of a tap.
- Safety tap prevents children from accessing single-faucet tub valves by covering them with a plastic box accessible to adults.
- Soft protective tub spout covers, usually in the shape of toy animals, protect when falling against the faucet's hard edge.
- Cushioned tub guards protect when falling against the edge of a metal tub.

of exposure at 120 degrees. Water heater tanks have adjustable thermostats for resetting the water temperature.

Filling up a tub or sink with the drain closed presents another dilemma. Either fixture can fill with water faster than the overflow protection can drain it.

Ease of exiting a wet tub isn't a concern only for the elderly. Install grab bars for kids to use when leaving the bath. Wall-mounted models and clamp-on styles are worth considering. It's the little things that will make the biggest impact in safety in your home.

YELLOW LIGHT

Some studies mention concern for bacterial growth if the water temperature is less than 140 degrees F, but this applies more to people with compromised immunization systems. Children have developing systems, and if this is a concern, discuss it with your pediatrician and balance the risks of burns versus possible water quality problems.

Grab Bars

- Grab bars or safety bars come in a variety of sizes and designs to accommodate people with different physical needs.
- Having a grab bar makes sense for anyone standing up in a slippery tub; install a long one so both adults and children can use it.
- To fasten to a tile wall, use a glass and tile drill bit for drilling the hole and stainless steel screws for securing.
- Hardware is available for installing inside a fiberglass or acrylic shower stall.

Garden Hoses

- Many hoses are made from PVC, which uses lead as a stabilizer; lead can also be found in brass hose fittings.
- Although lead levels have been reduced in recent years due to legal action against hose makers, some lead still remains and is noted on hose warning labels.
- A lead-free hose will be marked as "Safe for Drinking."
- Flush any hose out for a minute or two before drinking from it or filling a wading pool.

OTHER CHILD CONCERNS

Homes are built by and for adults—a little child orientation helps

With children, accidents happen, but the aim should be to prevent the mindless and traumatic accidents more than the bumps and bruises from normal play. Observe a child for a day, and you'll get a good idea what you have to lock up, limit access to, or ban from your kids.

Medicine looks like candy, and pediatric versions can taste appealing. It's little wonder children want to explore medicine cabinets. A good practice to keep pills out of the child's reach is locking medicine cabinets.

Safety gates on outside decks will keep toddlers from tumbling down the stairs, but decks that aren't built to current code can have openings in their railings large enough for a child to pass through them (openings between pickets or other vertical members should be 4 inches or less).

Space Heaters

- Both portable space heaters and hard-wired individual room heaters can get dangerously hot to the touch as well as be fire hazards.
- Select a space heater with a guard around the heating element or flame area to prevent children from getting burned.
- Look for a heater with sensors that turn a heater off when objects are too close or if children or pets move too near.
- Use heaters on the floor only and avoid hooking up with extension cords.

Video and Audio Monitoring

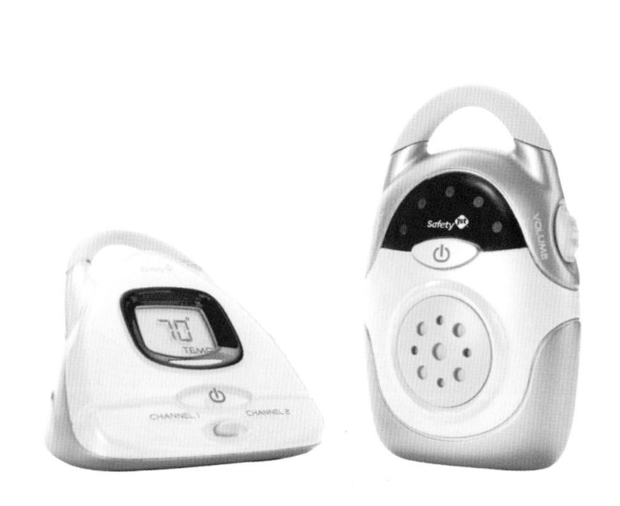

- Multipurpose baby monitors differ in built-in features, working range, and price.
- Some monitors feature two- and three-way receiving, while others offer video monitoring of two separate rooms.
- Monitor size varies from very portable—about the size of a home phone— to the size of a small TV set.
- Some monitors sound alarms if a child stops moving altogether for longer than twenty seconds, and others offer night-vision features.

Baby audio monitors aren't new, but the newest generation of audio/video monitors allows you to watch child activity from anywhere in the house using a monitor the size of a cell phone. These aren't a substitute for being in the same room, but like all safety devices, they're a good supplement. Making your home safe will keep the child out of trouble and ease your fears.

MAKE IT EASY

When time is at a premium, a home childproofer might be the way to go. A childproofer is a specialty contractor who does a home survey, provides an estimated cost for labor and safety materials, and does the installations. With online resources, you can do the same job if the contract cost is too high.

Cushion Edges for Furniture

- Kids can often get hurt by sharp furniture edges, particularly shorter children who are at eye level to these edges.
- Cushioning the edges with self-adhesive soft material protects without leaving marks on the furniture.
- No tools are required, and the cushions are removable when no longer needed.
- To prevent children from pulling over bookshelves and small cabinets, secure them to wall studs using angle brackets (a useful idea in areas of earthquakes, too).

Safety and Deck Railings

- Deck construction must meet local building code requirements and any homeowner association rules you might be subject to.
- Deck railings are a main safety feature, even for close-to-the-ground decks.
- A railing must be at least 36 inches tall and not have any gaps between any two vertical members more than 4 inches wide—this prevents babies from crawling through and falling.
- The railing must be exceptionally strong and well secured—lives can depend on it.

REFRIGERATOR

Even new, sophisticated refrigerators need some basic care and feeding

New refrigerators can cost thousands of dollars, depending on the model and features, which include ice and water dispensers, water filtration systems for these same water dispensers, multiple defrost options, and zoned cooling, and at least one model has a built-in TV that is satellite- and Internet-compatible.

Regardless of the age, style, or model of your refrigerator, it will need some attending to if you want it to run efficiently, nothing more than regular cleaning and inspection. As far as repairing water pumps or replacing a condenser fan, these can be exercises in frustration. You'll have to diagnose the problem, track down the part from an appliance store, and

Cleaning the Refrigerator

- A refrigerator's coils either are accessed through a front kick plate under the door or are located on the back of the refrigerator.
- Unplug the refrigerator or shut the power off, remove the kick plate, vacuum all the dust, or pull the refrigerator out and vacuum the rear coils.
- For dustier coils, use a coil-cleaning brush available at appliance parts stores.
- Remove the back plate and vacuum the dust from the condenser fan as well.

Checking Seals

- A refrigerator door gasket has to be kept intact, clean, and supple to remain an effective seal.
- Condensation on the seals suggests they are leaking and need to be replaced.
- Clean the seals monthly, after sticky spills, or when there's any presence of mold; use warm water and liquid dish soap (use a bleach-based cleaner for mold).
- When it is clean and dry, coat the seal with a small amount of petroleum jelly to keep it soft and prevent it drying out.

do the installation. And, while appliance repair technicians are expensive, they are often worth it. A job like this might be best left to a professional unless you have the time and patience to do it yourself.

ZOOM

To calculate whether a refrigerator is worth repairing, figure the cost of a service call, additional labor, parts, and the age of the appliance versus replacement with a unit under warranty that uses less energy. One repair can cost over $200, and a new basic refrigerator can cost over $500.

You might want to wear gloves while doing this process.

Cleaning

- There's no getting around regular refrigerator cleaning if you want to avoid bad food smells and unintentional rotten food.
- Unplug the refrigerator and remove everything, storing items that must be kept cold in a portable beverage cooler and tossing out anything that even looks questionable.
- Clean inside and out with warm water and liquid dish soap—use baking soda on stubborn spills.
- Rinse and dry all washed surfaces.

Replacing a Copper Water Line to the Ice-Maker

- Copper pipe is the highest quality material for most plumbing installations, but soft copper can be problematic for refrigerators with ice-makers.
- Soft, flexible 1/4-inch copper tubing is sometimes used to connect refrigerators to a cold water supply valve installed behind the refrigerator.
- Moving the refrigerator pulls on the tubing, which can become damaged.
- Replace the copper with a braided, flexible stainless steel connector, similar to what's recommended for washing machines.

DISHWASHER

While they are convenient and reliable, dishwashers do require some low maintenance

A study done at the University of Bonn is often quoted in the debate about dishwasher versus washing dishes by hand. The study found a dishwasher used less energy, water, and soap than hand-washing and got dishes cleaner. One study on its own rarely settles anything, but this one seems to be unchallenged.

A dishwasher can be especially efficient and energy-saving if the dishes are not excessively rinsed—or rinsed at all—before loading them, the dishwasher is full before running it, and the dry cycle is skipped in favor of air drying with the door open after the rinse cycles are finished. Also, skipping any rinse-hold and prerinse cycles will save additional energy and not affect cleaning the dishes.

Checking Gaskets

- Dishwasher door gaskets can harden and crack with age, allowing water to leak.
- Wipe the gasket clean once a month with a warm, soapy water solution, and when it is dry, rub on a small amount of petroleum jelly to keep it soft.
- Replace leaking gaskets; pull the damaged gasket out with a pair of pliers, and take it to an appliance parts store for a replacement.
- Soften the new gasket in warm water and install according to the manufacturer's instructions.

Rust on Dish Racks

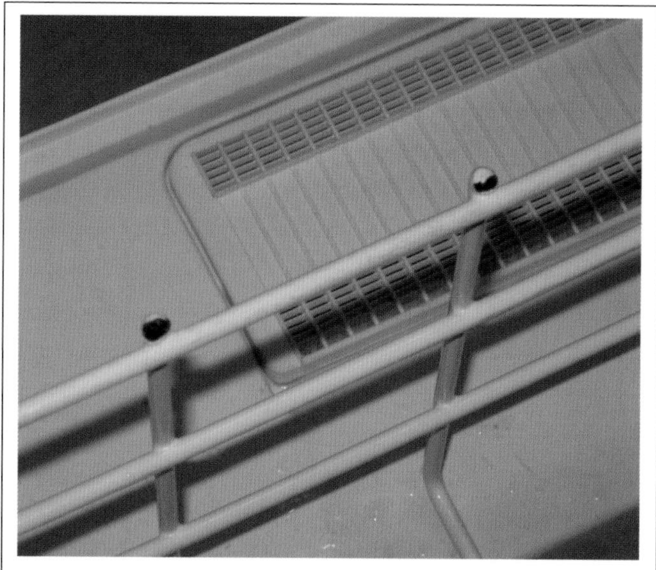

- When a dishwasher rack's vinyl coating wears out, the exposed sections of the metal rack will rust.
- Special brush-on sealants, available at appliance parts stores, offer limited partial repairs, as the sealants eventually wash away.
- Wrapping small strips of aluminum foil around the rust will protect dishes resting on those spots from rust stains.
- Compare the price of replacement racks with a new dishwasher—by the time they start rusting the dishwasher is old enough to need other repairs.

Dishwashers are wonderful appliances and can have a very long lifetime if you do some minimal routine maintenance. Like any motorized, wet environment, dishwashers wear down, especially ones that are used for more than one load a day, which isn't uncommon in family settings.

Dishwashers don't need much maintenance, but periodic inspections can help keep them running. With prices starting around $300 for a basic model, you have to evaluate whether it's worth repairing an older unit. You might find it's more cost-efficient to buy a newer, energy-saving model.

MAKE IT EASY

When replacing a dishwasher, consider whether the extra features, such as multiple cleaning cycles and electronic controls, are worth the money. These features can add to repair costs or the frequency of repairs, depending on the manufacturer and any history of warranty problems surrounding certain features or models.

Drains and Filters

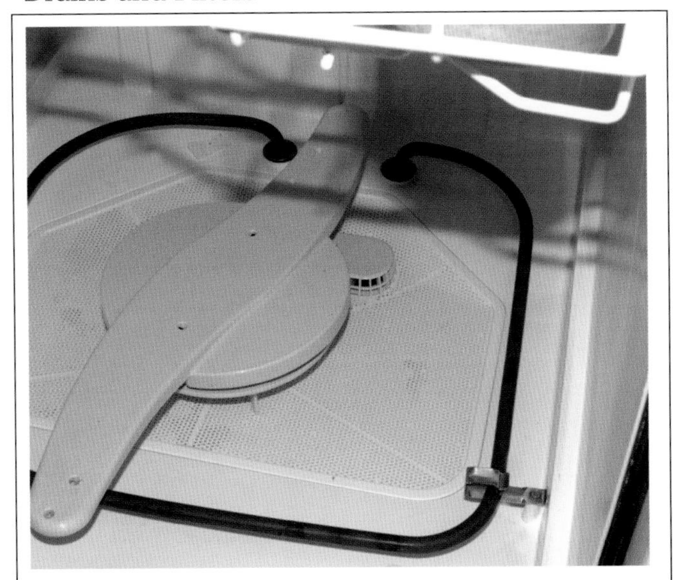

- Depending on the model and age of your dishwasher, you might have very little maintenance to do—check your owner's manual for recommendations.
- When a dishwasher fills slowly, check the water strainer, which is probably clogged.
- If it's draining slowly, look at the drain screen and remove any bits of food, broken glass, or other debris.
- Drain hoses can also get clogged—be sure to have a bucket ready when you loosen the end under the sink.

Water Sprayer

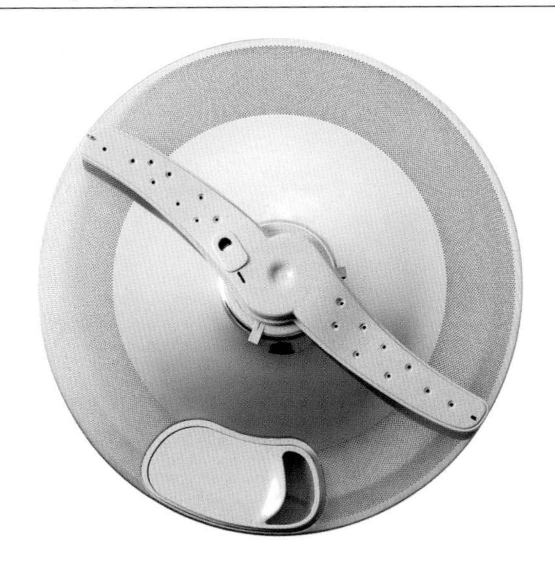

- Dishwasher water sprayers may become clogged with minerals from the water or detergent.
- Clean the water sprayer—including the upper sprayer, if present—as needed or every few months.
- Remove the unit and soak it in warm white vinegar to loosen mineral deposits.
- A spray arm is either secured with a bolt or is simply fitted over the pump and lifts off with a little bit of back-and-forth movement.

CLOTHES DRYER

For your safety, perform regular maintenance to make sure your dryer is up to speed

Before clothes dryers, homemakers hung clothes to dry, and it took forever. Today, clothes dryers are a huge convenience and are one of your home's biggest energy users. Like dishwashers, dryers should be run with full loads for most efficient use but not so full that the clothes don't easily tumble with sufficient air flow. Drying back-to-back loads of washed clothes saves some energy as well since the dryer doesn't have to completely warm up from a cold start.

Dryers exhaust lint from the clothes they dry. Lint clogs lint screens and exhaust hoses and needs to be regularly removed for safety (you want to avoid a lint fire) and efficiency. Dryer parts can eventually fail after years of use. Heating ele-

Lint Trap

- The U.S. Consumer Product Safety Commission estimates there are over fifteen thousand dryer-related fires every year.
- Clean the lint screen after every load and routinely vacuum out the lint trap itself.
- Vacuum behind the dryer once a month or so to get up any loose lint and other debris.
- Certain fabric softener sheets can build up a residue on the lint screen and clog it—be sure to wash the screen once every few weeks with warm water and soap.

Exhaust Hose

- Plastic dryer hoses should be replaced with either solid metal or flexible aluminum (gas dryers have other requirements).
- Do not attach the dryer hose to the exhaust ducting with metal screws—these can catch and trap lint.
- With your dryer running, check the vent outlet to be sure hot air is easily exhausted and the vent is not clogged with lint.
- Clean the hose and ducting with an electric blower from the inside after disconnecting the dryer hose from the dryer.

ments break in electric dryers, as do switches, fuses, and, in the worst case, motors. As with any major appliance repair, weigh replacement versus a visit from an appliance technician. Keep in mind that new dryers start at less than $300. Consider the dryer's age and repair history before deciding.

RED LIGHT

Never vent a dryer inside your home, as tempting as that might seem. Some argue it's all right, especially in dry climates where the humidity would be welcome, but the moisture would be excessive and damaging. Always vent a clothes dryer outside and keep the exhaust hose and vent cover clean of lint.

Heating Element

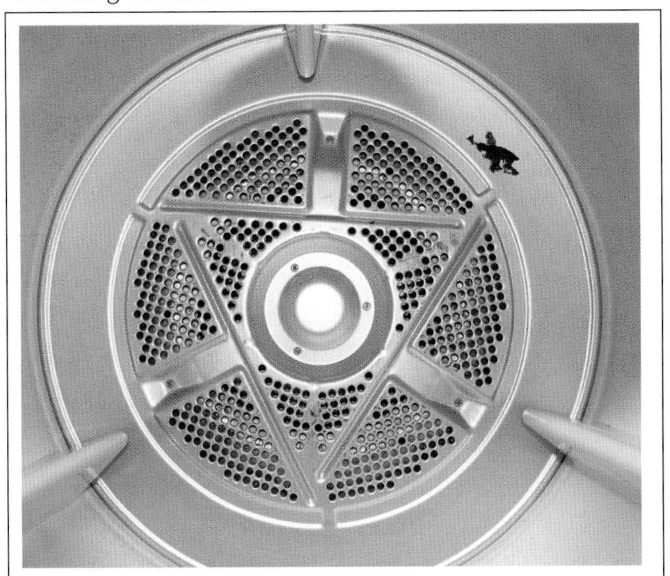

- When an electric dryer stops heating up, a regular source of the problem is the heating element, which is found at the back of the dryer.
- Heating elements are available at appliance part stores and are relatively inexpensive.
- Before removing the element, unplug the dryer—never work on an electrical component with the power connected.
- A damaged heating element will have burned-out, disconnected metal wire(s).

Ventilation

- If the interior of the drum is hot, but a normal-size load of clothes isn't dry after a long cycle, check that there is no lint obstruction anywhere.
- A kinked dryer hose can cause problems; it should be as straight as possible with the fewest bends necessary.
- Modern dryers have longer venting distances and ducting than in the past, making clean ducting more important than ever.
- Ideally, install a dryer directly against an exterior wall for a short venting distance.

GAS FURNACE

Keep your furnace working on the coldest days with some preparation before winter

The vast majority of home heating systems use natural gas furnaces. Gas is efficient, clean, and readily available. Before the dominance of natural gas, coal and later home-delivered oil were the main energy sources for home heat. Who wants to go back to having oil delivered or shoveling coal?

Gas furnaces are sold in two major categories: mid-efficiency (approximately 80 percent efficient) and high-efficiency (90–97 percent efficient). High-efficiency furnaces are more expensive than mid-efficiency models by approximately a thousand dollars or so and are apt to be a better investment in a cold climate than a moderate one. A high-efficiency furnace does require

Changing Filters

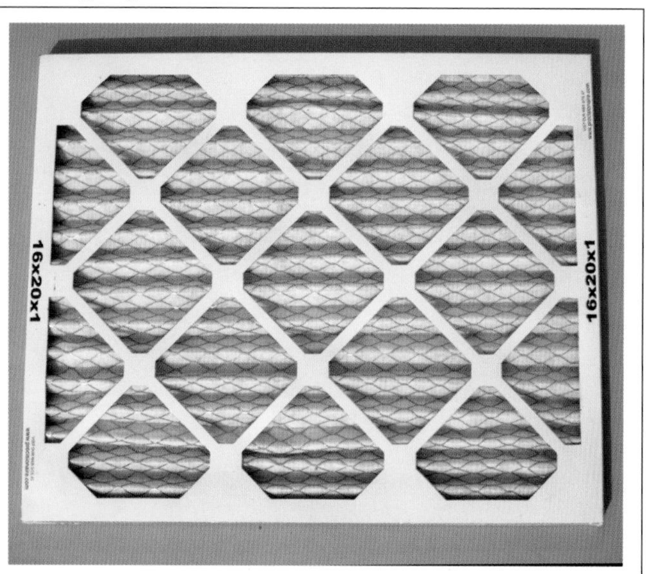

- Newer gas furnaces are far more efficient than older models, but they still require attention such as regularly changing the filters.
- Some filters are washable and can be reused, but they eventually tear and need replacement with new reusable filters or disposable filters.
- Besides general dust, shedding pets are another reason to change filters.
- Reusable filters range from basic fiberglass to costlier allergy, electrostatic, and HEPA-type filters.

Oiling the Furnace Motor

- Some older furnaces require their motors be lubricated with oil periodically—check your owner's manual.
- Do not attempt to lubricate a motor that doesn't require it.
- Look for a lubricating port accessible on the outside of the motor and add only the required amount of oil—usually just a few drops.
- Never use a spray lubricant such as WD-40 to lubricate a furnace motor; use only lubricating oil.

separate venting through a PVC pipe installed through the side of your house instead of using the chimney.

Is it worth going for the highest efficiency furnace? Some installers claim a high-efficiency furnace is more prone to repairs because it's more complicated than a mid-efficiency furnace.

Any furnace will work better with regular maintenance, including attending to the filters, having annual inspections by a heating technician, and, if needed, having professional duct cleaning.

Inspecting the Exhaust Ducting

- Both the exhaust ducting and the heat ducts should be intact, completely connected, and without any corrosion.
- Replace any corroded ducting—most likely it will be exhaust ducting.
- If there is any loose or broken cement sealer around the exhaust pipe/chimney connection, repair it before running the furnace.
- Check that the manual dampers in a forced-air system's ductwork are open and work—they balance the system's hot air flow.

ZOOM

New homes have programmable thermostats, but older homes often have mechanical, mercury-switch thermostats that function by reacting to the ambient temperature. A programmable thermostat controls heating and cooling by multiple time settings as well as temperature. A new thermostat must match the voltage rating (low-voltage, mini-voltage, or line voltage) of the old thermostat.

Schedule a Yearly Inspection

- A yearly inspection might seem excessive, especially if your furnace gets a clean bill of health, but think of it as an insurance policy.
- Use only a licensed and bonded heating contractor to perform this inspection.
- An inspection will test for lethal carbon monoxide leakage, a critical step since carbon monoxide is odorless and colorless.
- Expect to pay at least $100 for an inspection and service, depending on your location and available discounts.

WATER HEATER

Keeping tabs on your water heater's condition means fewer surprise cold showers

Your water heater provides you with warmth and stability and will do so for a longer period of time if you maintain it properly. According to the Federal Home Loan and Mortgage Corporation, you can expect a gas water heater to last up to twelve years. Routine maintenance will maximize its life.

Most water heaters are gas-powered. According to the EPA's ENERGY STAR program, heating water accounts for approximately 15 percent of a home's energy use. A high-efficiency water heater uses 10–50 percent less energy than a standard heater. Most water heaters have a storage tank, which is an insulated tank (30–80 gallons for the majority of homes) full of hot water. These tanks incur standby losses from energy

Checking the Pilot Light

- If a gas water heater stops providing hot water, check the pilot light.
- Gas water heaters have pilot lights, rather than electronic ignitions, to help maintain the tank's water temperature.
- When an electric water heater cools down, be sure the breaker hasn't tripped; check the high temperature cutoff in the water heater (open the panel and push the reset button).
- Follow the steps posted on the side of the tank when relighting a gas pilot light.

Other Checks

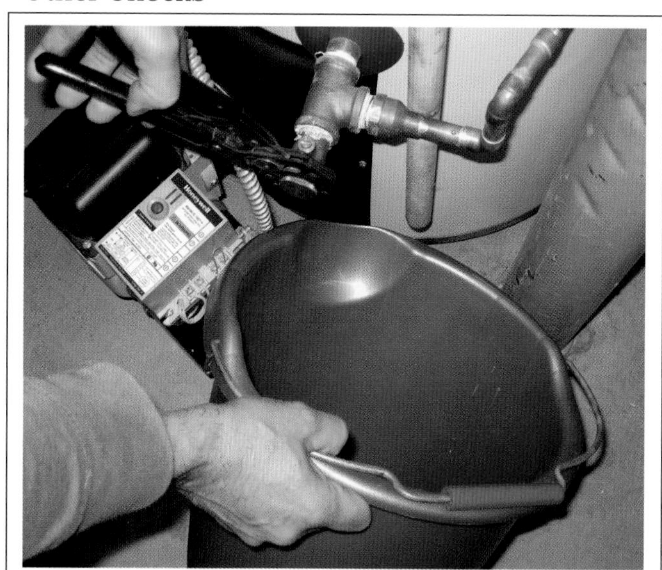

- Even minor maintenance should extend the life of a water tank.
- Emptying a few buckets full of water from the drain valve at the bottom of the tank will alert you to the presence of rust (suggests the anode rod needs replacement).
- Test the pressure release valve (a safety device) once a year, but be forewarned testing it might reveal the valve needs replacement.
- Electric heaters will take longer to heat up than gas models.

used to keep this water heated whether it's being used or not. Less popular on-demand, or tankless, water heaters heat water as it passes through a heating mechanism instead of storing it in a tank.

Sometimes a water heater will let you know the relationship is turning lukewarm, and you'll know it's time to step in and patch things up. Keep close tabs on the heater to see if repairs are needed. Some minor monitoring and maintenance will keep it going for years.

ZOOM

A water heater's temperature can be set as high as 160 degrees but is typically factory set at 120. Extremely hot water can scald. If your water heater has a "vacation" setting on its temperature knob, set it if you'll be away for five days or more to lower your energy bill.

Mineral Deposits

- Heating water—and hot water is more corrosive than cold—causes its naturally occurring minerals to settle faster and in larger quantities than those from unheated water.
- The sediment settles at the bottom of a water tank, where it can clog the drain valve and reduce the water tank's efficiency.
- Hard water contains more minerals and is more predominate than soft water.
- If too much scale builds up, the tank could fail.

Replacing the Anode Rod

- An anode rod attracts the minerals in the water that would otherwise accumulate on the bottom of the tank.
- Also called a sacrificial rod, it deteriorates in lieu of the tank.
- The anode rod should be replaced if the hot water rod is extensively corroded or smells bad.
- In tight spaces, replacing a full-length anode rod isn't always feasible, but a flexible type is available for these installations.

VACUUM CLEANER

Empty and check your cleaner regularly to maximize its cleaning capabilities

The first practical home vacuum cleaner was manufactured in the early 1900s by the Electric Suction Sweeper Company, which eventually became the Hoover Company. Vacuum cleaners took the drudgery out of carpet and rug cleaning. The invention of the disposable dust bag made life easier for those with allergies and made disposal faster.

Modern vacuum cleaners are light-years beyond earlier models. They're quieter, have stronger suction and larger dust and dirt storage, and come with attachments for cleaning furniture, window coverings, and stairs. After years of depending on bags for dirt disposal, the bagless design has gained a lot of traction, although emptying the

Emptying the Bag

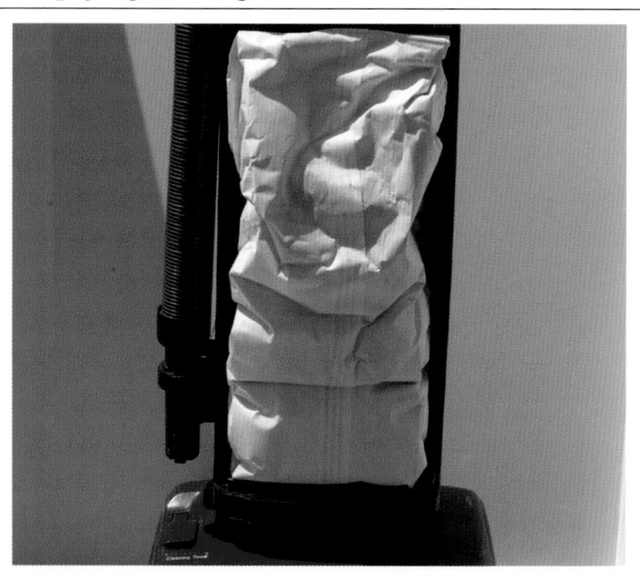

- Full bags cause the vacuum cleaner to work harder and vacuum less—empty the bag when the volume is up to the indicator line.
- It's a bit messy, but you can pull the contents out of a bag and reuse it if you're out of replacements.
- Increasingly, home-style vacuum cleaners are being replaced with bagless canister models.
- Bags come rated for standard filtration and high filtration for those with allergies.

Belts

- Belts need regular replacement, once a year or so with regular use—no tools required.
- Keep extra belts on hand in case of unexpected breakage.
- Avoid vacuuming up screws, pennies, and other small objects that can damage the fan or the motor.
- Cords can go bad from being repeatedly bent as they're wound up—replacing a cord can salvage a dead vacuum cleaner.

canister exposes the user to dust and debris.

Many a vacuum cleaner gets tossed out because it's clogged up with dust or the cord gets pinched and goes bad. If your cleaner is picking up less and straining more, it's time for some maintenance. Observe your cleaner's performance carefully to check for any inconsistencies. These inconsistencies will tip you off to any repairs needed. Regular cleanings will keep a vacuum cleaner working to full capacity without undue strain on the motor.

YELLOW LIGHT

Do your research before buying a bagless vacuum cleaner. One imported brand with extensive marketing, which has promoted the bagless design, is more expensive than well-rated domestic models. The difference can be a couple of hundred dollars—enough to buy a spare vacuum cleaner for cleaning up after repair jobs instead of using your good vacuum.

Brushes

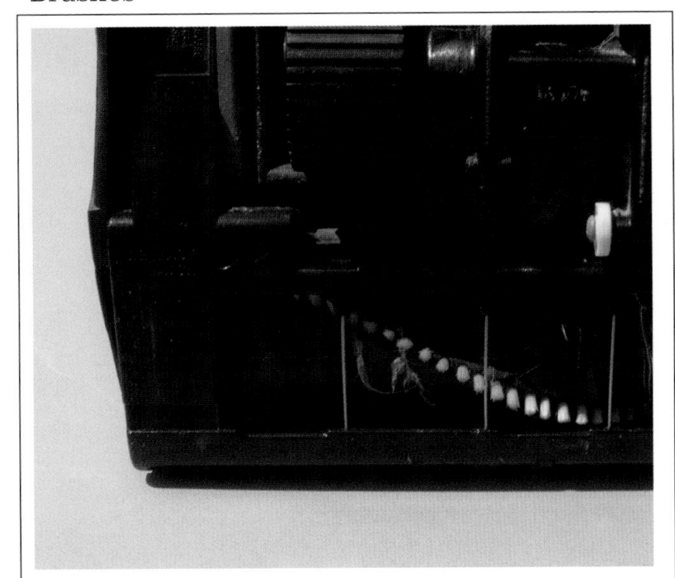

- A vacuum cleaner's brushes get tangled up with strings, rubber bands, and so on—cut these away to keep the brush roller moving freely.
- Whenever a vacuum cleaner suddenly stops operating, check the fuse or circuit breaker first.
- Although upright vacuums are more popular, the advantage of a canister model is a larger motor can be available for vacuuming.
- Replace or clean any built-in filters according to the manufacturer's recommendations.

Electric Blower

- Electric blowers are highly versatile tools that extend the life of other tools and appliances as well as perform standard clean-up chores.
- Use a blower to clean sawdust from drills, sanders, and saws.
- The best way to clean out vacuum cleaner hoses, brushes, and canisters is using a blower.
- When gutters are dry, blow the leaves and twigs out instead of scooping.

SUMMER

Make this the time for vacations, barbecues, and the inescapable warm weather repairs

Let's face it: If we could skip home maintenance, we would, especially during the summer. When a sunny weekend comes around, who wants to be working on the house? No one, but you don't want to be regretting it when the weather turns in the fall and certainly not in the winter when it's far too cold to touch up the paint or repair a broken glass pane. Pace yourself so you don't get buried by the jobs and let the summer pass you by completely.

Start with the outside. This is the time to do any siding and roof repairs while it's dry and you're not fighting wet weather conditions. You might find out after removing trashed siding that the damage is more extensive than you thought and

Paint Exterior

- Touch up any areas where the paint is missing, flaking, or bubbling.
- Paint entire lengths of siding or whole sections when touching up so the newly painted area doesn't stick out too much.
- If the entire house needs painting, schedule carefully and pace yourself so you're done in a timely manner without giving up all your free time or consider calling a painting contractor.
- Get your painting done by early September, even if you expect a mild autumn.

Window and Siding Repairs

- Damaged or rotten siding can be covering a larger problem underneath—start these repairs early in the season.
- Prime both sides of any replacement siding and prepaint the first coat before installation.
- Window repair can mean removing a sash for more than a day—be prepared to secure the opening with plywood.
- Be sure to paint any new glazing compound; otherwise, it will dry up and eventually fall out.

requires removing the sub-siding as well.

Summer is the obvious time for painting, but it also has its limitations. Too much heat or high humidity affects both interior and exterior paints. Painting in hot sunlight can cause the paint surface to dry too fast and form a skin before the rest of the coating is completely dry, resulting in pinholes, blisters, and excessive brush marks in the finish. As with any project, set up a timeline for getting the work done.

MAKE IT EASY

In hot weather, paint won't flow and level properly straight out of the can, but it can be diluted, as much as 10 percent, if necessary for easier application. It's tempting to paint late into the evening, but overnight dew might affect the paint gloss. Stop your painting about two hours before sunset to be safe.

Deck, Roof, and Gutter Repairs

- If you're up on the roof replacing damaged shingles, inspect the surrounding area and replace any questionable shingles as well.
- The less walking around you do on an old roof, the better since just being up there can damage shingles.
- After gutter repair and cleaning, run a hose to check your work and look for additional leaks.
- Recoat your deck for extra protection, even if there isn't a pressing need for it.

Filters

- Replace or clean your air conditioning filters at the start of the season and then every one or two months thereafter—more often if the summer is dry and dusty.
- Clean or replace furnace filters as well.
- Routinely check on and get your air conditioning system serviced.
- Summer is the best time for furnace servicing—if a part has to be ordered, you won't go without heat since the furnace isn't in use.

FALL

Take advantage of the cooler (yet not cold) weather to do some maintenance around the house

Fall is sort of a batten down the hatches time in many parts of the country. Winter cold and wet are coming, the warm weather is heading south, and daylight is slowly disappearing. All those lovely colorful leaves are falling and too many of them end up in your gutters. Ignore them and your gutters are likely to overflow in wet weather. The overflow can loosen paint on the siding and the retained water weighs down the gutters, straining their fasteners and joints. It's time to bring in the garden hoses and drain the outside pipes and hose bibs to avoid freezing and splitting.

Do you have removable window screens? Removing and storing them allows you to wash the windows one more

Screens and Storm Windows

- As soon as the bugs are gone, remove your window screens, wash them, and store them until spring.
- While the screens are down, wash the windows, inside and out.
- If you have combined screen/storm window units, lubricate the tracks as you change the screens for storm windows and wash the outside of the storm windows.
- Install any removable storm windows, checking that they're clean first.

Chimney and Gutters

- Inspect your chimney and clean it if that wasn't done earlier in the year.
- Be sure the masonry joints are in good shape and consider installing a chimney cap if you don't already have one.
- Clean the gutters thoroughly and be prepared to do it again once all the leaves are down.
- In cold climates, shut off the water to all outside hose bibs, drain them at the faucets, and put the hoses away in a garage or basement.

time before it turns freezing outside.

In cold climates, we all spend more time indoors around the wood stove or fireplace. You don't want to take a chance on a chimney fire or an electrical fire for that matter. Fall is a great time for an annual safety check of all systems, including water and gas shut-offs and smoke detectors.

YELLOW LIGHT

Carbon monoxide poisoning from furnaces and other sources of incomplete combustion contributes to more than two thousand deaths annually, according to the *Journal of the American Medical Association*. A standard furnace inspection should include checking carbon monoxide levels. Consider a carbon monoxide detector, which sounds an alarm when detecting unsafe levels of carbon monoxide.

Odds and Ends

- To avoid loosening their wires, tighten any receptacles that have cords regularly inserted and pulled out (mostly in the kitchen and the bathrooms).
- Replace any burned-out light bulbs in outside lighting and check that motion detectors are properly aimed.
- Check that the main water shut-off is functioning and that it moves easily.
- Replace all your smoke detector batteries, checking to see if your community has a battery collection or recycling program, and test the detectors.

Safety Inspection

- With winter coming, you can expect to spend more time indoors with lights and heating systems running longer hours plus using a fireplace or wood-burning stove.
- Be sure all family members know the locations of gas, water, and electricity shut-offs and how to use them.
- Lay out an evacuation plan in the event of a fire.
- Clean out the dryer exhaust duct and replace any kitchen exhaust fan filters.

WINTER

Winter's here, so start up some routine cold weather maintenance

If the snow is falling, it's a little late to paint the windows or clean the gutters, so here's hoping all that maintenance is finished. Winter maintenance chores are limited, but there are still some pesky tasks to attend to.

Do you have air conditioning? Snow, ice, and pine needles don't do your air conditioning units any favors. Cover them up for the winter and keep corrosion to a minimum. Furnaces are heavily used in the winter and call for monthly filter changes. If your furnace hasn't been inspected in a few years and you missed it during the past summer, call a technician now and be prepared to wait. This is the busy season for heating contractors, but you want your furnace checked before it dies on Christmas Eve.

Replace Furnace Filter Monthly

- Replace or clean your furnace filter monthly during the heating season and check that a gas furnace flame is blue and burning at a consistent rate—a yellow, unsteady flame requires a technician immediately.
- If you don't feel sufficient heat coming out of a register and you have flexible ducting running in a crawl space, check that the ducting is still tight to the register.
- Repair or throw away any damaged holiday lights.
- Have one operable window per bedroom for emergency escapes.

Car Check

- Your car works harder in cold weather, and a breakdown in the snow can strand you.
- Be sure your car has been serviced for winter driving with an antifreeze check, proper temperature-rated windshield fluid, and a healthy battery.
- Carry an extra scraper and battery cables.
- In severe winter climates, carry road salt or Kitty Litter for traction, a tow rope, and possibly a small shovel for digging out of snow.

Are you prepared to be housebound for a day or two or three in heavy snow, possibly without power? Be a prepared homeowner and have nonperishable food, water, flashlights, and candles stored away.

Cold winter driving is hazardous and demanding on car batteries. Do you have a winter survival kit in your car trunk? Getting stuck in a blizzard can happen to anyone, and it's not difficult to prepare for it. Like with any other season, being prepared for winter weather will keep your home and life running smoothly.

YELLOW LIGHT

Wood-burning fireplaces present possible hazards. Creosote build-up in masonry and metal chimneys can eventually catch fire. Hot fires can weaken chimney mortar. The Chimney Safety Institute recommends homeowners who have three or more winter fires a week should have their chimneys inspected and cleaned once a year.

Ice Dams

- When ice dams form at the edges of roofs, they can cause melting snow to back up and leak into the attic space.
- Carefully clearing off snow with a snow rake or push broom will temporarily ease the problem.
- Do not aggressively attack ice dams with pounding tools—there's too much risk of damaging the roof and causing a leak.
- When pipes freeze, open the taps and use a hair dryer, not a torch, to thaw them out.

Preparing for an Electrical Outage

- Cold weather and impassable roads compound electrical outages in the winter.
- Store some nonperishable food and bottled water, waterless hand cleaner, a flashlight, candles, manual can opener, and battery-operated radio or TV in the event of weather-related electrical outage.
- Place candles inside metal bowls, baking dishes, or even sinks for safe long-term, unattended burning.
- Keep a similar kit inside your car for roadside emergencies.

SPRING

Spring brings the yearly maintenance cycle full circle—it's time to start again

You think of spring, and spring cleaning comes to mind, a fresh start now that the snow and cold are gone. You may do your own repairs or pay to have them done, but there's no way around maintenance if you want a safe, appealing place that lives up to basic expectations.

A regularly maintained house means you don't get hit with much bigger deferred problems later, so be mindful of routine check-ups. Spending a weekend in the spring on a chore list is a small price to pay to kick off a new, warmer season.

First, inspect the outside of your house for winter wear and tear. Give it a bath while you're at it and wash off any soot, grime, or other seasonal gunk. Take your storm windows

Prep for Warmer Weather

- Remove, wash, and store any removable storm windows and install the window screens and storm door screens.
- Repair any windows that you want open and operable—get them now before the weather turns too hot.
- Remove the air conditioner cover and check for leaks, then clean it, allow it to fully dry, and store it.
- Remove the picnic table cover or, if there was no cover, see if the table needs recoating and schedule this in.

Cleaning Up Winter Messes

- The winter leaves grunge, and the spring is a good time to renew the outside with a scrub and wash.
- As soon as your soil starts to dry out, rake out any decomposing leaves and twigs from the garden beds.
- Clean out any fireplace ashes, screens, grates, and glass doors and, if present, shut off the gas pilot light as soon as the weather is too warm for fires.
- Clean and store all winter-related items, including clothing.

down, put screens up, and turn outside water on. Thinking about painting in the summer? Start considering the amount of preparation, color selection, and budget for the project now. Call for some painting bids if you're hiring it out before painters get booked for the season. Do the same with roofing contractors if your roof needs replacing.

And don't forget to throw the windows open, air your home out, and enjoy the spring.

ZOOM

During the days of wood- and coal-burning stoves, spring cleaning was necessary to clean a winter's worth of accumulated soot. Modern heating systems have eliminated this source of grime, but spring is still viewed as a time to renew and refresh a home after being closed up all winter.

Exterior Checks

- Check the roof for winter damage as soon as possible.
- Check the attic space for leaks and wet spots on the rafters or insulation as well as the roof shingles and flashing for damage.
- Masonry mortar can require repair after going through cycles of freezing and thawing—look for sandy texture, cracks between rows of bricks, and missing chunks of mortar.
- Confirm your foundation is dry without any puddles and the gutters are intact.

Odds and Ends

- Turn water on at hose bibs and check for breaks in the pipes.
- If any hoses were left out all winter, connect them to the water heater tank and run hot water through them to soften them some.
- If any spots in your yard remain notably wet, investigate installing additional drainage.
- Winter is especially hard on automobiles—detail your car inside and out to remove oxidized paint and thoroughly clean the carpets.

MONTHLY

Make monthly maintenance a regular event like cooking or doing laundry

Life is full of repetition: sleeping, shopping for groceries, eating big holiday meals and regretting it later. To keep your maintenance and repairs to a manageable level, treat them the same way. Instead of waiting for problems to appear, try to prevent them.

As despised as "to-do" lists are, they're better than "absolutely must do now before the floor caves in" lists later. We accept that dishes have to be washed daily and clothes washed once or twice a week so accepting that smoke detectors should be checked monthly isn't much of an imaginative stretch.

Having trouble following a regular maintenance list? Set up a schedule and post it on the refrigerator or as a timed

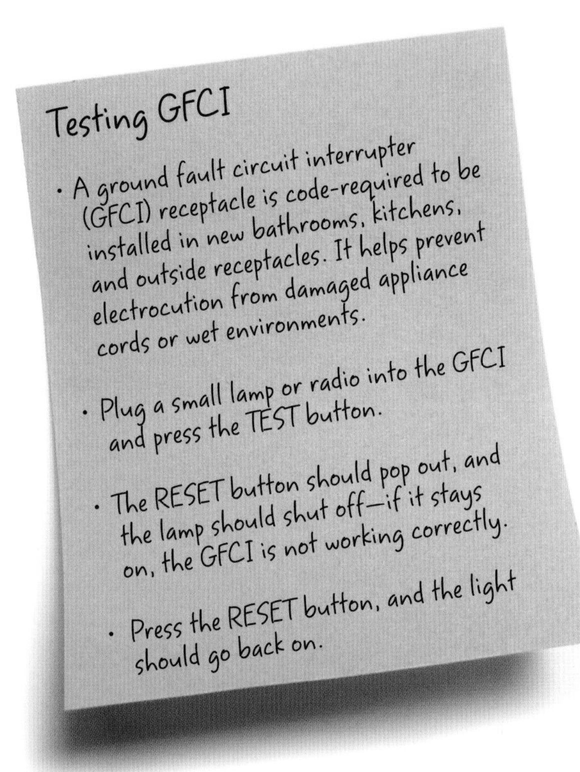

Check Smoke Alarms

- FEMA and other government agencies state that smoke detectors should be tested once a month by pressing the test button (do not test with an open flame).
- If the alarm fails to sound off, replace it immediately.
- When a smoke detector begins randomly beeping, it usually means the batteries need replacement.
- Smoke alarms have a useful life of about ten years, at which point they should be replaced, even if they appear to be working.

reminder on your computer or wireless device. This visual reminder will help you set up a monthly routine. Make a note to check the list every month, take care of the task, and be done with it. No single task is especially onerous and is certainly easier than dealing with the consequences of not attending to it.

YELLOW LIGHT

Pouring boiling water down drains is a great way to keep them clear, but water this hot can crack molded one-piece sink/vanity top combinations. Use a funnel inserted into these drains to pour boiling water. Never pour boiling water into a toilet as there is some chance it could crack the bowl.

Drains

- Every drain has a trap to hold water, which acts as a barrier against sewer gases coming up through the drain lines.
- Pour water down unused or rarely used drains once a month to compensate for evaporation.
- In general, maintain drains with a baking soda and vinegar mix followed by hot water to flush it down.
- Grind ice cubes and a lemon wedge in the food disposer to help eliminate odors and clean away sludge.

Filters

- Change or clean the furnace/air conditioning filters once a month, depending on use.
- Clean the kitchen exhaust metal mesh fan filters (put them in the dishwasher or let them soak).
- Inspect and clean faucet aerators and shower heads—remove any rust particles and soak aerators and shower heads in vinegar for a few hours to break up any hard water deposits.
- If your refrigerator has a removable drain pan, remove and clean it according to the manufacturer's recommendations and vacuum the coils.

YEARLY

Once a year isn't much to ask for these important jobs

Some jobs come around only once a year—preparing taxes and cleaning up after New Year's Eve come to mind—and they vary in importance and how much time they'll consume. If you use your wood-burning chimney each fall and winter, inspecting it once a year—not just for creosote but also the raccoons and birds that sometimes set up home in chimneys—is important.

Yearly jobs tend to get done during decent weather, some in preparation for colder, wetter weather to come and others because after a year of use, dust or lint or moisture can accumulate in places you don't want it, and it behooves you to take care of them.

If you live alone, you are stuck with doing this yourself or hiring out. Otherwise, get your partner and kids involved.

Odds and Ends

- Vacuum the dust out of smoke detectors for safer operation and from heating registers and radiators for better heat distribution.
- If you've never had your heat ducts cleaned, consider having this professionally done.
- "Bleeding" your radiator system once a year is a process of removing trapped air that decreases the system's efficiency.
- Your computer needs dusting, too—back up your data, take the PC outside, open up the case, and blow out the inside with a can of compressed air.

Clean and Seal Tile Grout

- Grout is a long-lasting material, but it cannot stand up indefinitely to water in shower areas.
- Seal tile grout once a year following the manufacturer's directions for application and drying times.
- If any grout deteriorated, remove it with a grout saw or similar tool, remove any loose caulking, regrout, and then seal all the tile.
- After grouting, apply new caulk rated for tub and tile use.

Once they're old enough, children can learn to check plumbing valves and the condition of bathroom tile the same as an adult. Whether these lessons and responsibilities stick will remain to be seen, but at least you share the chore list, and this in itself is a good lesson.

And, no, you probably don't want your eight-year-old climbing around on the roof, but taking an electric blower and blasting the dust out of the garage will get any child's attention.

MAKE IT EASY

Schedule any yearly chore that might involve a contractor following up early enough that any needed work can be done before the contractor's busy season. If it's painting, you have time to talk with more than one painter and might even get a better estimated price by booking early.

Inspect Chimney

- Wood-burning fireplaces and wood stoves create creosote, which can build up to dangerous levels inside chimneys.
- To inspect these chimneys, start from the roof and check the condition of the masonry and then look down the flue with a strong flashlight or lower a work light down with a rope.
- Check for creosote build-up (it will appear as a shiny glaze on the chimney walls).
- If you suspect damage or creosote, call a chimney specialist.

Service Furnace

- Yearly furnace servicing might seem unnecessary, but it's inexpensive insurance, especially as a furnace ages.
- An inspection should cover function, air testing, the condition of the fire box, burners, drafting, and carbon monoxide levels.
- Check that all plumbing shut-off valves at every fixture and the main shut-off still function properly and inspect toilets for stability.
- Look in the basement, crawl space, and attic for pest and rodent infestation.

RESOURCES

When it comes to information on home repair and maintenance, there are plenty of websites to turn to for further reading or manufacturer suggestions. Although this list is by no means exhaustive, it will get you started on other sources of information for home repairs and maintenance, whether it's looking for the right tools, proper safety techniques, or alternate home maintenance options.

Chapter 1: Problem Solving

Plumbing Leaks:
www.homeandgardenadvice.com

Repairing Pipes:
www.howstuffworks.com

Gas Leaks:
www.hannabery.com

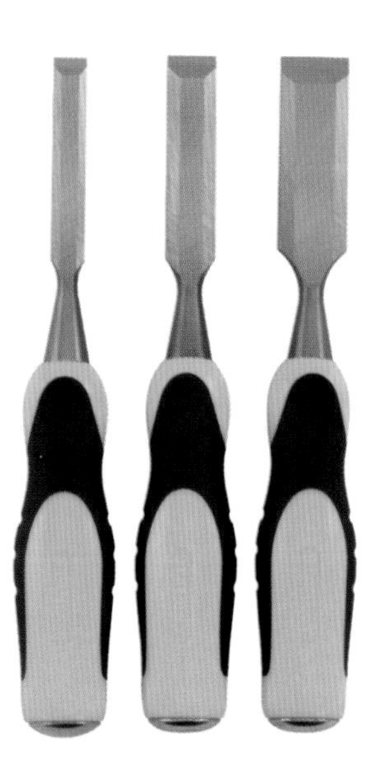

Home Security:
www.homesecurityinformation.com

The Federal Alliance for Safe Homes:
www.flash.org

Chapter 2: Vital Tools

Tool Safety and Ergonomics:
www.cdc.gov/niosh

Hammers:
www.hammernet.com

Screwdrivers:
www.doityourself.com/screwdrivers

Pliers:
www.hardwarestore.com

Wrenches:
www.acehardware.com

Pry Bars:
www.toolsofthetrade.net

Japanese Saws:
www.japanwoodworker.com

Scrapers:
www.jamestowndistributors.com

Drills:
www.homeenvy.com

Electric Sanders:
www.doityourself.com/sanders

Measuring Tape:
www.perfecttape.com

Wood Chisels:
www.sawdustmaking.com

Nail Sets:
www.lowes.com

Pressure Washers:
www.ultimatewasher.com

Chainsaws:
www.stihlusa.com/chainsaws

Other:
www.boschtools.com
www.stanleyworks.com
www.wagnerspraytech.com

Chapter 3: Vital Hardware

Screws:
www.nutsandbolts.com

Bolts:
www.boltscience.com

Nails:
www.diydata.com

Caulk:
www.dap.com

Glue:
home.howstuffworks.com

Sandpaper:
www.abrasivesoasis.com

Hearing Protection:
www.cdc.gov/niosh

Respiratory Protection:
www.cdc.gov/niosh

Information about Lead-Based Paint:
www.epa.gov/lead

Chapter 4: Vital Products

Paint, Primer, Sheens, & Glosses:
www.behr.com
www.benjaminmoore.com
www.sherwin-williams.com

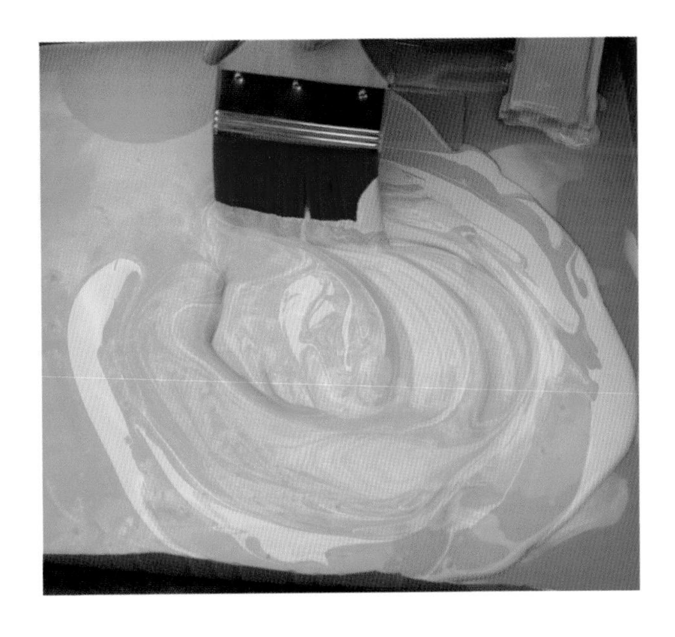

www.vansicklepaint.com
www.paint.org

Other Wood Finishes:
www.hardwarestore.com

Shellac:
www.zinsser.com
www.woodworker.com

Recycled Paint:
www.metropaint.org
www.dunnedwards.com
www.visionsrecycling.com
www.boomerangpaint.com
www.ecospaints.com
www.hotzenvironmental.com/paint
www.recyclepaint.com
www.localcolorinc.com

Automotive Body Filler:
www.naturalhandyman.com

Wood Filler:
www.dap.com

Chapter 5: Scratches & Chips

Refinishing:
www.refinishfurniture.com

Wood Floor Repairs:
www.woodfloordoctor.com

Plastic Laminate:
www.formica.com

Tips for Fixing Cracked Tile:
www.hometime.com

Tips for Grouting Tile:
www.hometime.com

Chapter 6: Inside Leaks/Clogs

Quick Fixes for Leaky Faucets:
www.doityourself.com
www.deltafaucet.com

Tips for Repairing Toilets:
www.acehardware.com

Fixing a Leaking Washing Machine:
www.repairclinic.com

Chapter 7: Outside Leaks/Clogs

Gutter Covers:
www.leaffilter.com
www.waterloov.com

Repairing Leaks:
www.repair-home.com

Windows:
www.andersenwindows.com

Chapter 8: Squeaks/ Sticky Issues

Fixing Floor Squeaks:
www.hometips.com
www.bobvila.com

Fixing Stair Squeaks:
www.bobvila.com

Fixing Sticking Doors:
www.lowes.com
www.repair-home.com

Tape for Sticky Drawers:
www.tool-warehouse.com

Drawer Guides:
www.slidedummy.info

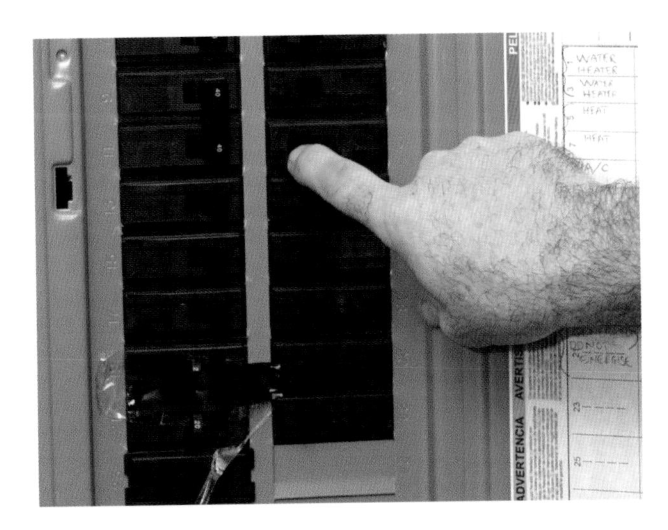

Door Lock Repairs:
www.hometips.com

Sticking Drawers:
www.ronhazelton.com

Window Repairs:
www.windowrepair.com

Chapter 9: Cracks & Holes

Plaster Damage Repairs:
www.homedepotmoving.com

Plaster Crack Repairs:
www.hometips.com

Drywall Holes:
www.askthebuilder.com
www.epotek.com

Large Drywall Repairs:
www.easy2diy.com

Drywall Cracks:
www.acehardware.com

Concrete Cracks:
www.concretenetwork.com

Cracked Glass:
www.cornerhardware.com
www.gtglass.com

Chapter 10: Electrical Systems

Electrical System Information:
www.nfpa.org
www.cpsc.gov
www.nahb.org
www.nationalgridus.com
www.ge.com

Fuses and Circuit Breakers:
www.pge.com
www.fuses.cc

Circuit Breakers:
www.pge.com
www.nationalswitchgear.com

Fuses:
www.fuses.cc

Resetting a Breaker:
www.ehow.com

Chapter 11: Fungi Issues

Mold and Mildew:
www.epa.gov/mold/moldresources
www.bradlewis.com

Wood Rot & Repairs:
www.hammerzone.com
www.ewoodcare.com

Chapter 12: Pets & Pests

Installing a Pet Door:
www.lowes.com
www.cornerhardware.com
www.gundoghousedoor.com

Raccoons and Squirrels:
www.popularmechanics.com
www.aaanimalcontrol.com

Bees and Wasps:
www.pestworld.org
www.orkin.com

Birds:

www.birdbgone.com

Extermination:
www.pestworld.org
www.doyourownpestcontrol.com
www.terminix.com

Other:
www.rubbermaid.com

Chapter 13: Painting

All about Paint:

www.paint.org
www.sherwin-williams.com
www.benjaminmoore.com
www.behr.com
www.glidden.com

Wood Finishes:
www.woodzone.com
www.woodfinishsupply.com

How to Paint: www.paintquality.com

Exterior Preparation:
www.cornerhardware.com

Chapter 14: Vital Storage

Building Shelves:
www.askthebuilder.com
www.ikea.com

Closet Storage:

www.easyclosets.com
www.closetorganizersusa.com
www.stacksandstacks.com

Garage Shelving:

www.tidygarage.com
www.carguygarage.com
www.garagetek.com

Workbench:
www.workbench-ideas.com

Chapter 15: Energy Efficiency

Energy Losses:
www.dsireusa.org

Sealing Your Home:
www.energystar.gov

Weather Stripping:
www.eere.energy.gov

Energy Efficient Appliances:
www.energystar.gov
www.kohler.com

Chapter 16: Outside Repairs

Fence Repairs:
www.fenceportal.com
North American One-Call Referral Center: 888.258.0808

Compost:
www.compostguide.com
www.epa.gov

Outdoor Lighting:
www.acehardware.com
www.besthomeledlighting.com
www.electricsuppliesonline.com
www.ylighting.com

Replacing Deck Boards:
www.decks.com
www.trex.com
www.correctdeck.com

Rain Barrels:
www.rainbarrelguide.com

Chapter 17: Childproofing

Childproofing:
www.cpsc.gov
www.nachi.org
www.safety1st.com
www.totsafe.com

Chapter 18: Appliances

Appliance Maintenance:
www.repairclinic.com
www.applianceaid.com

Chapter 19: Timeline Maintenance

www.nahb.org
www.bobvila.com
www.statefarm.com
www.homeinspectorlocator.com

GLOSSARY

A/C: An abbreviation for air conditioner or air conditioning.

Aerator: The round screened screw-on tip of a sink spout. It mixes water and air for a smooth flow.

Aggregate: A mixture of sand and stone, and a major component of concrete.

Anchor bolts: Bolts to secure a wooden sill plate to concrete, or masonry floor or wall.

Attic access: An opening that is placed in the drywalled ceiling of a home providing access to the attic.

Attic ventilators: In houses, screened openings provided to ventilate an attic space.

Backing: Frame lumber installed between the wall studs to give additional support for drywall or an interior trim related item, such as handrail brackets, cabinets, and towel bars. In this way, items are screwed and mounted into solid wood rather than weak drywall that may allow the item to break loose from the wall.

Balusters- Vertical members in a railing used between a top rail and bottom rail or the stair treads. Sometimes referred to as pickets or spindles.

Base or baseboard: A trim board placed against the wall around the room next to the floor.

Base shoe: Molding used next to the floor on interior base board. Sometimes called a carpet strip.

Batt: A section of fiber-glass or rock-wool insulation measuring 15 or 23 inches wide by 4 to 8 feet long and various thicknesses.

Beam: A structural member transversely supporting a load. A structural member carrying building loads from one support to another.

Bearing wall: A wall that supports any vertical load in addition to its own weight.

Blankets: Fiber-glass or rock-wool insulation that comes in long rolls 15 or 23 inches wide.

Blow insulation: Fiber insulation in loose form and used to insulate attics and existing walls where framing members are not exposed.

Brace: An inclined piece of framing lumber applied to wall or floor to strengthen the structure. Often used on walls as temporary bracing until framing has been completed.

Breaker panel: The electrical box that distributes electric power entering the home to each branch circuit (each plug and switch) and composed of circuit breakers.

Building codes: Community ordinances governing the manner in which a home may be constructed or modified.

Cap: The upper member of a column, pilaster, door cornice, molding, or fireplace.

Casement: Frames of wood or metal enclosing part (or all) of a window sash that may be opened by means of hinges affixed to the vertical edges.

Casement window: A window with hinges on one of the vertical sides and swings open like a normal door.

Casing: Wood trim molding installed around a door or window opening.

Caulk: A flexible material used to seal a gap between two surfaces e.g., between pieces of siding or the corners in tub walls.

Ceiling joist: One of a series of parallel framing members used to support ceiling loads and supported in turn by larger beams, girders or bearing walls. Also called roof joists.

Ceramic tile: A man-made or machine-made clay tile used to finish a floor or wall. Generally used in bathtub and shower enclosures and on countertops.

Circuit: The path of electrical flow from a power source through an outlet and back to ground.

Circuit breaker: A device which looks like a switch and is usually located inside the electrical breaker panel or circuit breaker box. It shuts off the power the house and limits the amount of power flowing through a circuit (measured in amperes).

Condensation: Beads or drops of water that accumulate on the inside of the exterior covering of a building.

Conductivity: The rate at which heat is transmitted through a material.

Conduit: A pipe, usually metal, in which wire is installed.

Contractor: A company licensed to perform certain types of construction activities.

Counter flashing: A metal flashing used on chimneys at the roofline to cover shingle flashing and used to prevent moisture entry.

Crawl space: A shallow space below the living quarters of a house, normally enclosed by the foundation wall and having a dirt floor.

Crown molding: A molding used on cornice or wherever an interior angle is to be covered, especially at the roof and wall corner.

Damper: A metal "door" placed within the fireplace chimney. It is normally closed when the fireplace is not in use.

Dead bolt: An exterior security lock installed on exterior entry doors that can be activated only with a key or thumb-turn. Unlike a latch, which has a beveled tongue, dead bolts have square ends.

Door jamb: The surrounding case into which and out of which a door closes and opens. It consists of two upright pieces, called side jambs, and a horizontal head jamb. These three jambs have a door stop installed on them.

Door stop: The wooden style that the door slab will rest upon when it's in a closed position.

Dormer: An opening in a sloping roof, the framing of which projects out to form a vertical wall suitable for windows or other openings.

Double glass: Window or door in which two panes of glass are used with a sealed air space between. It's also known as insulating glass.

Double hung window: A window with two vertically sliding sashes, both of which can move up and down.

Downspout: A pipe, usually of metal, for carrying rainwater down from the roof's horizontal gutters.

Drywall: A manufactured panel, commonly 1/2-inch thick and 4X8' or 4X12' in size, made out of gypsum plaster and encased in heavy paper. The panels are nailed or screwed onto the framing and the joints are taped and covered with a joint compound.

Ducts: The heating system. Usually round or rectangular metal pipes installed for distributing warm (or cold) air from the furnace to rooms in the home. Also a tunnel made of galvanized metal or rigid fiberglass, which carries air from the heater or ventilation opening to the rooms in a building.

Face nail: To install nails into the vertical face of a bearing header or beam.

Fire retardant chemical: A chemical or preparation of chemicals used to reduce the flammability of a material or to retard the spread of flame.

Flashing: Sheet metal or other material used in roof and wall construction to protect a building from water seepage.

Flat paint: An interior paint that contains a high proportion of pigment and dries to a flat or lusterless finish.

Foundation: The supporting portion of a structure below the first floor construction, or below grade, including the footings.

Fuse: A device often found in older homes designed to prevent overloads in electrical lines. This protects against fire.

GFCI or GFI: (Ground Fault Circuit Interrupter) An ultra sensitive plug designed to shut off all electric current. Used in bathrooms, kitchens, exterior waterproof outlets, garage outlets, and wet areas.

General contractor: A contractor who enters into a contract with the owner of a project for the construction of the project and who takes full responsibility for its completion, although the contractor may enter into subcontracts with others for the performance of specific parts or phases of the project.

Grain: The direction, size, arrangement, appearance, or quality of the fibers in wood.

Grout: A wet mixture of cement, sand, and water that flows into masonry or ceramic crevices to seal the cracks between the different pieces.

Gutter: A shallow channel or conduit of metal or wood set below and along the eaves of a house to catch and carry off rainwater from the roof.

Gypsum plaster: Gypsum formulated to be used with the addition of sand and water for base-coat plaster.

Hardware: All of the metal fittings that go into the home when it is near completion.

Humidifier: An appliance normally attached to the furnace, or portable unit device designed to increase the humidity within a room or a house by means of the discharge of water vapor.

HVAC: An abbreviation for heat, ventilation, and air conditioning.

I-beam: A steel beam with a cross section resembling the letter I. It is used for long spans as basement beams or over wide wall openings, such as a double garage door, when wall and roof loads bear down on the opening.

I-joist: Manufactured structural building component resembling the letter I. Used as floor joists and rafters.

Insulation: Any material high in resistance to heat transmission that, when placed in the walls, ceiling, or floors of a structure, will reduce the rate of heat flow.

Jamb: The side and head lining of a doorway, window, or other opening. It includes studs as well as the frame and trim.

Joint: The location between the touching surfaces of two members or components joined and held together by nails, glue, cement, mortar, or other means.

Joint compound: A powder that is usually mixed with water and used for joint treatment in gypsum-wallboard finish. It is often called spackle or drywall mud.

Joist: Wooden 2X8's, 10's, or 12's that run parallel to one another and support a floor or ceiling, and supported in turn by larger beams, girders, or bearing walls.

Kilowatt (kw): One thousand watts. A kilowatt hour is the base unit used in measuring electrical consumption.

Lattice: An open framework of criss-crossed wood or metal strips that form regular, patterned spaces.

Masonry: Stone, brick, concrete, hollow-tile, concrete block, or other similar building units or materials.

Molding: A wood strip having an engraved, decorative surface.

NEC (National Electrical Code): A set of rules governing safe wiring methods. Local codes, which are backed by law, may differ from the NEC in some ways.

Nozzle: The part of a heating system that sprays the fuel of fuel-air mixture into the combustion chamber.

Panel: A thin flat piece of wood, plywood, or similar material, framed by stiles and rails as in a door (or cabinet door), or fitted into grooves of thicker material with molded edges for decorative wall treatment.

Particleboard: Plywood substitute made of coarse sawdust that is mixed with resin and pressed into sheets.

Pilot hole: A small-diameter, pre-drilled hole that guides a nail or screw.

Pilot light: A small, continuous flame (in a hot water heater, boiler, or furnace) that ignites gas or oil burners when needed.

Plywood: A panel (normally 4 X 8') of wood made of three or more layers of veneer, compressed and joined with glue, and usually laid with the grain of adjoining plies at right angles to give the sheet strength.

Post: A vertical framing member usually designed to carry a beam. Often a 4x4", a 6x6", or a metal pipe with a flat plate on top and bottom.

Pressure Relief Valve (PRV): A device mounted on a hot water heater or boiler which is designed to release any high steam pressure in the tank to prevent tank explosions.

Primer: The first, base coat of paint when a paint job consists of two or more coats. A first coating formulated to seal raw surfaces and holding succeeding finish coats.

Putty: A type of dough used in sealing glass in the sash, filling small holes and crevices in wood, and for similar purposes.

PVC or CPVC: (Poly Vinyl Chloride) A type of white or light gray plastic pipe sometimes used for water supply lines and waste pipe.

Radiant heating: A method of heating, usually consisting of a forced hot water system with pipes placed in the floor, wall, or ceiling.

Rafter: Lumber used to support the roof sheeting and roof loads. Generally, 2X10's and 2X12's are used. The rafters of a flat roof are sometimes called roof joists.

Receptacle: An electrical outlet.

Sash: A single light frame containing one or more lights of glass. It is the frame that holds the glass in a window, often the movable part of the window.

Sealer: A finishing material, either clear or pigmented, that is usually applied directly over raw wood for the purpose of sealing the wood surface.

Semigloss paint or enamel: A paint or enamel made so that its coating, when dry, has some luster but is not very glossy. Bathrooms and kitchens are normally painted semi-gloss.

Service entrance panel: Main power cabinet where electricity enters a home wiring system.

Shim: A small piece of scrap lumber or shingle, usually wedge shaped, which when forced behind a furring strip or framing member forces it into position.

Shingles: Roof covering of asphalt. asbestos, wood, tile, slate, or other material cut to stock lengths, widths, and thickness.

Shingles, siding: Various kinds of shingles, used over sheathing for exterior wall covering of a structure.

Short circuit: A situation that occurs when hot and neutral wires come in contact with each other. Fuses and circuit breakers protect against fire that could result from a short.

Siding: The finished exterior covering of the outside walls of a frame building.

Single hung window: A window with one vertically sliding sash or window vent.

Soffit: The area below the eaves and overhangs.

Strike: The plate on a door frame that engages a latch or dead bolt.

Stucco: An outside plaster finish made with Portland cement as its base.

Stud: A vertical wood framing member, also referred to as a wall stud, attached to the horizontal sole plate below and the top plate above.

Subfloor: The framing components of a floor to include the sill plate, floor joists, and deck sheeting over which a finish floor is to be laid.

Taping: The process of covering drywall joints with paper tape and joint compound.

Termites: Wood eating insects that superficially resemble ants in size and general appearance, and live in colonies.

Terra cotta: A ceramic material molded into masonry units.

Toenailing: To drive a nail in at a slant. Method used to secure floor joists to the plate.

Trap: A plumbing fitting that holds water to prevent air, gas, and vermin from backing up into a fixture.

Tread: The walking surface board in a stairway on which the foot is placed.

Veneer: Extremely thin sheets of wood.

Vent: A pipe or duct which allows the flow of air and gasses to the outside. Also, another word for the moving glass part of a window sash, i.e., window vent.

Voltage: A measure of electrical potential. Most homes are wired with 110 and 220 volt lines. The 110 volt power is used for lighting and most of the other circuits. The 220 volt power is usually used for the kitchen range, hot water heater, and dryer.

Weatherstrip: Narrow sections of thin metal or other material installed to prevent the infiltration of air and moisture around windows and doors.

PHOTO CREDITS

Chapter 1
vii (left): © Vincent Giordano/shutterstock
vii (right): © photos.com
1 (left): © George Peters/istockphoto
2 (right): © Amy Walters | Dreamstime.com
3 (left): © Mark Evans/istockphoto
4 (left): © VisualField/istockphoto
4 (right): © Ryan Klos/istockphoto
5 (left): Marilyn Zelinsky-Syarto
6 (right): Marilyn Zelinsky-Syarto
7 (right): © Norman Pogson/istockphoto
8 (left): © Jim Lopes/shutterstock
8 (right): © Gautier Willaume/istockphoto
9 (left): © Jonathan Lenz/shutterstock
9 (right): © Olivier Le Queinec/ shutterstock
10 (left): Marilyn Zelinsky-Syarto
10 (right): Marilyn Zelinsky-Syarto
11 (left): Anna Adesanya

Chapter 2
12 (right): Courtesy of Bosch
13 (right): © Gillian Mowbray/istockphoto
14 (left): Courtesy of The Stanley Works
14 (right): Courtesy of The Stanley Works
15 (left): Courtesy of The Stanley Works
15 (right): Courtesy of The Stanley Works
16 (left): Courtesy of The Stanley Works
16 (right): Courtesy of The Stanley Works
17 (left): Courtesy of The Stanley Works
17 (right): Courtesy of The Stanley Works
18 (left): Courtesy of Bosch
18 (right): Courtesy of Bosch
19 (left): Courtesy of The Stanley Works
19 (right): Courtesy of The Stanley Works
20 (left): Courtesy of Bosch
20 (right): Courtesy of Bosch
21 (left): Courtesy of Bosch
21 (right): Courtesy of Bosch
22 (left): Courtesy of Wagner
22 (right): Anna Adesanya
23 (left): Courtesy of Bosch
23 (right): © Kingjon | Dreamstime.com

Chapter 3
24 (left): Anna Adesanya
24 (right): Anna Adesanya
25 (left): Anna Adesanya
25 (right): Anna Adesanya
26 (left): Anna Adesanya
26 (right): Anna Adesanya
27 (left): Anna Adesanya
27 (right): Anna Adesanya
28 (right): Anna Adesanya
29 (left): Anna Adesanya
30 (left): Anna Adesanya
30 (right): Anna Adesanya
31 (left): Anna Adesanya
32 (left): Anna Adesanya
33 (left): Anna Adesanya
33 (right): Anna Adesanya
34 (left): Anna Adesanya
34 (right): Anna Adesanya
35 (left): Anna Adesanya
35 (right): Anna Adesanya

Chapter 4
36 (left): Anna Adesanya
36 (right): © Dewayne Flowers | Dreamstime.com
37 (left): Anna Adesanya
38 (left): Anna Adesanya
38 (right): Anna Adesanya
39 (left): Anna Adesanya
40 (left): Anna Adesanya
40 (right): Anna Adesanya
41 (left): Anna Adesanya
42 (right): Anna Adesanya
43 (left): © photos.com
44 (left): © Misty Diller | Dreamstime.com
44 (right): © Peter Galbraith | Dreamstime.com
45 (left): Anna Adesanya
45 (right): Anna Adesanya
46 (left): Anna Adesanya
46 (right): Anna Adesanya
47 (left): Anna Adesanya
47 (right): Anna Adesanya

Chapter 5
48 (left): Anna Adesanya
48 (right): Anna Adesanya
49 (left): Anna Adesanya
49 (right): Anna Adesanya
50 (left): Anna Adesanya
50 (right): Anna Adesanya
51 (left): Anna Adesanya
51 (right): Anna Adesanya
52 (left): Anna Adesanya
52 (right): Anna Adesanya
53 (left): Anna Adesanya
53 (right): Anna Adesanya
54 (left): Anna Adesanya
54 (right): Anna Adesanya
55 (left): © Elenat | Dreamstime.com
55 (right): © Yvanovich | Dreamstime.com
56 (left): Anna Adesanya
56 (right): © Jon McIntosh/istockphoto
57 (left): © Christina Richards/istockphoto
58 (left): © Stephen R. Syarto
58 (right): Anna Adesanya
59 (left): © Christopher Hudson/ shutterstock
59 (right): © Robert Redelowski/ shutterstock

Chapter 6
60 (left): Courtesy of The Delta Faucet Company
60 (right): Anna Adesanya
61 (left): Anna Adesanya
61 (right): Marilyn Zelinsky-Syarto
62 (left): Jack Tom
63 (right): Jack Tom
63 (left): Jack Tom
64 (right): Anna Adesanya
65 (left): Anna Adesanya
66 (right): Marilyn Zelinsky-Syarto
67 (left): Jack Tom
67 (right): Anna Adesanya
68 (right): Anna Adesanya
69 (left): Jack Tom

Chapter 7
70 (left): Marilyn Zelinsky-Syarto
70 (right): Marilyn Zelinsky-Syarto
71 (left): © Marc Pinter | Dreamstime.com
71 (right): Anna Adesanya
72 (left): Stephen R. Syarto
72 (right): Anna Adesanya
73 (left): Marilyn Zelinsky-Syarto
73 (right): Anna Adesanya
74 (left): Anna Adesanya
74 (right): Jack Tom
75 (left): Anna Adesanya
76 (left): Courtesy of Anderson Windows
76 (right): Anna Adesanya
77 (right): © David Lewis/istockphoto
78 (left): Anna Adesanya
79 (left): Anna Adesanya
79 (right): Anna Adesanya
80 (right): Anna Adesanya
81 (left): Anna Adesanya
81 (right): Anna Adesanya

Chapter 8
82 (right): © David Park | Dreamstime.com
83 (left): © Carli Schultz Kruse/istockphoto
83 (right): Stephen R. Syarto
84 (left): Jack Tom
84 (right): Jack Tom
85 (left): Jack Tom
86 (right): Anna Adesanya
87 (left): Anna Adesanya
88 (left): Stephen R. Syarto
88 (right): Marilyn Zelinsky-Syarto
89 (left): © VanDenEsker/istockphoto
89 (right): Stephen R. Syarto
90 (left): Anna Adesanya
90 (right): Anna Adesanya
91 (left): Courtesy of Andersen Windows
91 (right): Marilyn Zelinsky-Syarto

Chapter 9
92 (left): © robcocquyt/shutterstock
92 (right): Marilyn Zelinsky-Syarto
93 (left): Marilyn Zelinsky-Syarto
93 (right): Marilyn Zelinsky-Syarto
94 (left): Anna Adesanya
94 (right): Anna Adesanya
95 (left): Anna Adesanya
95 (right): Anna Adesanya
96 (left): Maureen Graney
96 (right): Maureen Graney
97 (left): Maureen Graney
97 (right): Maureen Graney
98 (left): Jack Tom
98 (right): Jack Tom
99 (left): Maureen Graney
99 (right): Maureen Graney
100 (left): Anna Adesanya
100 (right): Anna Adesanya
101 (left): Anna Adesanya
101 (right): Anna Adesanya
102 (left): © Joe Klune/istockphoto
102 (right): Jack Tom
103 (left): Jack Tom
103 (right): Marilyn Zelinsky-Syarto

Chapter 10
104 (left): Marilyn Zelinsky-Syarto
104 (right): Marilyn Zelinsky-Syarto
105 (left): Marilyn Zelinsky-Syarto
106 (left): Anna Adesanya
106 (right): © photos.com
107 (left): © John Wollwerth/shutterstock
108 (right): © Lisa F. Young
109 (right): © Laura Scudder/Wikipedia/Wiki-media Commons
110 (left): Jack Tom
110 (right): Jack Tom
111 (right): Stephen R. Syarto
112 (right): Courtesy of General Electric
113 (left): Anna Adesa nya
114 (left): © Tom Tomczyk/istockphoto
114 (right): Courtesy of General Electric
115 (left): Courtesy of Energy Star

Chapter 11
116 (left): Anna Adesanya
116 (right): Anna Adesanya
117 (left): Anna Adesanya
118 (right): Courtesy of General Electric
120 (left): Anna Adesanya
120 (right): Marilyn Zelinsky-Syarto
121 (left): Marilyn Zelinsky-Syarto
121 (right): © Paul Senyszyn/istockphoto
122 (right): Marilyn Zelinsky-Syarto
123 (left): Anna Adesanya
123 (right): Marilyn Zelinsky-Syarto
124 (right): Stephen R. Syarto
125 (left): Marilyn Zelinsky-Syarto
125 (right): Marilyn Zelinsky-Syarto
126 (left): © Suzanne Paul | Dreamstime.com
126 (right): © Suzanne Paul | Dreamstime.com
127 (right): © Frances Twitty/istockphoto

Chapter 12
128 (left): Courtesy of Bradley J. Lewis @ bradlewis.com
128 (right): Marilyn Zelinsky-Syarto
129 (left): © Podius | Dreamstime.com
130 (left): © Digitoll | Dreamstime.com

131 (left): © Lein de León Yong/shutterstock
131 (right): © Rick Sargeant | Dreamstime.com
133 (left): Courtesy of GunDogHouse
133 (right): Marilyn Zelinsky-Syarto
134 (right): © Carolyn Seelen | Dreamstime.com
135 (left): © Lorna | Dreamstime.com
136 (left): Courtesy of Joshua Jones/A All Animal Control of Tri-State @ www.aallanimalcontrol.com
137 (left): © Robeo | Dreamstime.com
137 (right): Courtesy of Bird-B-Gone, Inc.@www.birdbgone.com
138 (left): © Ed Endicott / Wysiwyg Foto LLC/shutterstock
138 (right): © TAOLMOR/shutterstock
139 (left): © Christopher S. Howeth/shutterstock
139 (right): © Frank Boellmann/Shutterstock
140 (right): © Bonnie Watton/Shutterstock
141 (left): © Bonnie Watton/Shutterstock

Chapter 13
142 (left): Anna Adesanya
142 (right): Marilyn Zelinsky-Syarto
143 (left): Anna Adesanya
143 (right): Courtesy of Rubbermaid
144 (left): Courtesy of Glidden
144 (right): Anna Adesanya
145 (left): © Juriah Mosin/shutterstock
146 (left): Anna Adesanya
146 (right): Anna Adesanya
147 (left): Anna Adesanya
147 (right): Anna Adesanya
148 (right): Courtesy of Glidden
149 (left): Anna Adesanya
150 (left): Anna Adesanya
150 (right): Marilyn Zelinsky-Syarto
151 (left): © Wendy Kaveney | Dreamstime.com
151 (right): Courtesy of Glidden
152 (left): © Jack Schiffer | Dreamstime.com
152 (right): © Davinci | Dreamstime.com
153 (left): Anna Adesanya

Chapter 14
154 (left): © Frances Twitty/istockphoto
154 (right): Courtesy of Wagner
155 (left): © Lisa F. Young/Shutterstock
156 (right): Courtesy of Ikea
157 (left): Courtesy of Ikea
158 (right): Courtesy of Ikea
159 (left): Stephen R. Syarto
159 (right): © Ene | Dreamstime.com
160 (right): © MaxFX/shutterstock
161 (left): © John Wollwerth | Dreamstime.com
161 (right): Courtesy of Ikea
162 (left): Courtesy of GarageTek
162 (right): Courtesy of GarageTek
163 (left): Courtesy of GarageTek
163 (right): Courtesy of GarageTek
164 (right): © Kristin Smith | Dreamstime.com
165 (left): Stephen R. Syarto

Chapter 15
166 (right): Courtesy of Ikea
167 (left): Courtesy of Ikea
168 (right): Jack Tom
169 (left): Stephen R. Syarto
169 (right): © Nicholas Moore/Shutterstock
170 (left): Jack Tom
170 (right): © Limeyrunner | Dreamstime.com
171 (left): Anna Adesanya
171 (right): © Jeffery Stone/Shutterstock
172 (left): Anna Adesanya
173 (left): Anna Adesanya
173 (right): Anna Adesanya
174 (left): Anna Adesanya
174 (right): Marilyn Zelinsky-Syarto
175 (left): © Lastdays1 | Dreamstime.com
176 (right): © Kelpfish | Dreamstime.com
177 (left): © Kelpfish | Dreamstime.com
177 (right): © Sabryna Washington | Dreamstime.com

Chapter 16
178 (left): © Greenstockcreative | Dreamstime.com
178 (right): Anna Adesanya
179 (left): Courtesy of Kohler
179 (right): Anna Adesanya
180 (left): Claire Desjardins/istockphoto
180 (right): Marilyn Zelinsky-Syarto
181 (left): Marilyn Zelinsky-Syarto
181 (right): Marilyn Zelinsky-Syarto
182 (left): Marilyn Zelinsky-Syarto
182 (right): © Lastdays1 | Dreamstime.com
183 (left): Marilyn Zelinsky-Syarto
183 (right): Crystalcraig | Dreamstime.com
184 (left): Marilyn Zelinsky-Syarto
184 (right): Stephen R. Syarto
185 (right): Anna Adesanya
186 (left): Anna Adesanya
186 (right): Anna Adesanya
187 (left): Anna Adesanya
187 (right): © Darleen Stry | Dreamstime.com
188 (right): © Astarfotograf/istockphoto
189 (right): © PhillDanze/istockphoto

Chapter 17
190 (right): © Cornelia Pithart | Dreamstime.com
191 (right): © copit/shutterstock
192 (right): © Dorel Juvenile Group 2008. All Rights Reserved.
193 (left): Courtesy of Safe Beginnings Inc.
194 (left): Anna Adesanya
194 (right): © Diane Diederich/shutterstock
195 (left): Anna Adesanya
195 (right): © Dorel Juvenile Group 2008. All Rights Reserved.
196 (left): © Dorel Juvenile Group 2008. All Rights Reserved.
196 (right): © Dorel Juvenile Group 2008. All Rights Reserved.
197 (left): © tiburonstudios/istockphoto
197 (right): Marilyn Zelinsky-Syarto
198 (left): Courtesy of Ikea
198 (right): Anna Adesanya
199 (left): © Dorel Juvenile Group 2008. All Rights Reserved.
199 (right): © Dorel Juvenile Group 2008. All Rights Reserved.
200 (left): © Dorel Juvenile Group 2008. All Rights Reserved.
200 (right): © Don Bayley/istockphoto

201 (left): © Sorsillo | Dreamstime.com
201 (right): Anna Adesanya

Chapter 18

202 (left): © Martin Green | Dreamstime.com
202 (right): © Dorel Juvenile Group 2008. All Rights Reserved.
203 (left): © Dorel Juvenile Group 2008. All Rights Reserved.
204 (right): Courtesy of FisherPaykel
205 (left): Courtesy of Fisher Paykel
206 (left): Anna Adesanya
206 (right): Marilyn Zelinsky-Syarto
207 (left): Anna Adesanya
207 (right): Courtesy of Fisher Paykel
208 (left): Courtesy of Fisher Paykel
208 (right): Marilyn Zelinsky-Syarto
209 (left): © Kevin Norris/Shutterstock
210 (left): Anna Adesanya
210 (right): Marilyn Zelinsky-Syarto
211 (left): Anna Adesanya
212 (left): Anna Adesanya
212 (right): Marilyn Zelinsky-Syarto

Chapter 19

214 (left): Anna Adesanya
214 (right): Anna Adesanya
215 (left): Anna Adesanya
215 (right): Marilyn Zelinsky-Syarto
216 (left): Anna Adesanya
216 (right): Anna Adesanya
217 (left): Anna Adesanya
217 (right): Marilyn Zelinsky-Syarto
218 (left): Anna Adesanya
218 (right): Anna Adesanya
219 (left): Anna Adesanya
220 (left): Anna Adesanya
220 (right): Anna Adesanya
221 (left): Jack Tom
222 (right): Marilyn Zelinsky-Syarto
223 (left): Anna Adesanya
224 (right): Anna Adesanya
225 (left): Anna Adesanya
225 (right): Anna Adesanya
226 (right): Marilyn Zelinsky-Syarto
227 (left): Anna Adesanya
227 (right): Anna Adesanya

INDEX